AS/A-LEVEL YEAR 1

STUDENT GUIDE

AQA

Physics

Sections 1, 2 and 3
Measurements and their errors
Particles and radiation
Waves

Jeremy Pollard

PHILIP ALLAN FOR
HODDER
EDUCATION
AN HACHETTE UK COMPA

Philip Allan, an imprint of Hodder Education, an Hachette UK company, Blenheim Court, George Street, Banbury, Oxfordshire OX16 5BH

Orders

Bookpoint Ltd, 130 Milton Park, Abingdon, Oxfordshire OX14 4SB

tel: 01235 827827

fax: 01235 400401

e-mail: education@bookpoint.co.uk

Lines are open 9.00 a.m.–5.00 p.m., Monday to Saturday, with a 24-hour message answering service. You can also order through the Hodder Education website: www.hoddereducation.co.uk

© Jeremy Pollard 2015

ISBN 978-1-4718-4374-7

First printed 2015

Impression number 5 4 3 2 1

Year 2018 2017 2016 2015

This guide has been written specifically to support students preparing for the AQA AS and A level Physics (Sections 1, 2 and 3) examinations. The content has been neither approved nor endorsed by AQA and remains the sole responsibility of the author.

Cover photo: Beboy/Fotolia

Typeset by Integra Software Services Pvt Ltd, Pondicherry, India

Printed in Italy

Hachette UK's policy is to use papers that are natural, renewable and recyclable products and made from wood grown in sustainable forests. The logging and manufacturing processes are expected to conform to the environmental regulations of the country of origin.

Contents

Content Guidance

Questions & Answers

Getting the most from this book

Exam tips

Advice on key points in the text to help you learn and recall content, avoid pitfalls, and polish your exam technique in order to boost your grade.

Knowledge check

Rapid-fire questions throughout the Content Guidance section to check your understanding.

Knowledge check answers

1 Turn to the back of the book for the Knowledge check answers.

Summaries

■ Each core topic is rounded off by a bullet-list summary for quick-check reference of what you need to know.

Exam-style questions

Commentary on the questions

Tips on what you need to do to gain full marks, indicated by the icon **e**

Sample student answers

Practise the questions, then look at the student answers that follow.

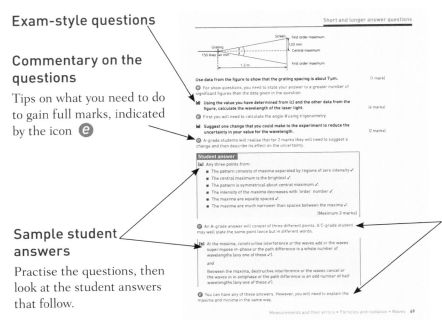

Commentary on sample student answers

Find out how many marks each answer would be awarded in the exam and then read the comments (preceded by the icon **e**) showing exactly how and where marks are gained or lost.

■About this book

This guide covers the first three sections of the A-level and AS specification for AQA Physics. It is intended to help you to remember and understand the physics you need for the course and is set out in the same order as the specification so you can check that you have covered everything.

There are two main sections:

■ The **Content Guidance** covers the main points of the topics. The detail is intended to help you understand what is needed and how to use that understanding and knowledge in questions. The worked examples should help you to understand the principles and to see the sorts of questions that you might be required to undertake in an examination. There are some quick knowledge check questions that will help you to be sure you understand each point of physics. There is an activity for each of the two required practicals. Answers are given at the end of the book.

■ The **Questions & Answers** section comprises two test papers with answers, so that you can practise questions and see the sorts of answers that are needed and the knowledge and understanding that is required. If you are taking an AS examination in physics, the questions will not be quite as difficult as the questions you will get on the same material for A-level. A-level questions will be a bit more synoptic, i.e. they will ask sub-questions on other parts of the specification not covered in this guide.

You will need to learn the basic facts and ensure you understand the connections between different ideas. It is often helpful to learn beyond the specification so that these connections become more obvious. The more you can do that, the better you will be able to tackle new questions or different ideas. If you try all the questions in this book and more besides, you will be able to approach any examination with confidence.

Content Guidance

■ Measurements and their errors

In **Measurements and their errors** you need to know about the use of SI units and their prefixes and the limitations and uncertainties associated with physical measurements. You also need to be able to describe the order of magnitude of quantities and to estimate measurements.

Use of SI units and their prefixes

Fundamental quantities and SI units 24/4/16 6/1/17

The seven fundamental (base) quantities and their SI units (abbreviated from the French Système International d'Unités) are the quantities that are used to derive all other quantities (and units). The fundamental quantities and their SI units are:

- length, metre (m)
- mass, kilogram (kg)
- time, second (s)
- current, ampere (A)
- temperature, kelvin (K)
- quantity of matter, mole (mol)
- light intensity, candela (cd)

Derived units

All non-fundamental quantities (and their units), such as speed, are derived in terms of the seven fundamental quantities. Speed, which is defined as distance (or a length) divided by time, has the units $m\,s^{-1}$. Some other quantities, such as force, are given a derived unit (normally named after a famous scientist — in this case the newton (N), where $1\,N = 1\,kg\,m\,s^{-2}$). Table 1 shows some of the more common derived units.

Exam tip

You will not be asked questions involving the quantity of light intensity and its unit, the candela, and you do not need to write a definition for each of the fundamental quantities in the examination.

You must always include the units of all calculated quantities in the examination.

A fundamental (base) quantity is a physical quantity that is used to derive other quantities and their units.

Knowledge check 1

What are the seven fundamental (base) units?

Quantity	Definition (formula)	Derived unit name and symbol	Base units
Frequency	$f = \dfrac{1}{t}$ ✓	Hertz, Hz ✓	s^{-1} ✓
Speed	$v = \dfrac{\Delta s}{\Delta t}$ ✓		$m\,s^{-1}$ ✓
Acceleration	$a = \dfrac{\Delta v}{\Delta t}$ ✓		$m\,s^{-2}$ ✓
Force	$F = m \times a$ ✓	Newton, N ✓	$kg\,m\,s^{-2}$ ✓
Pressure	$P = \dfrac{F}{A}$ ✓	Pascal, Pa ✓	$kg\,m^{-1}\,s^{-2}$ ✓
Work done (energy)	$W = \boxed{Fs\cos\theta}$	Joule, J ✓	$kg\,m^2\,s^{-2}$ ✓
Power	$P = \dfrac{\Delta W}{\Delta t}$ ✓	Watt, W ✓	$kg\,m^2\,s^{-3}$ ✓
Charge	$\Delta Q = I\Delta t$ ✓	Coulomb, C ✓	$A\,s$ ✓
Potential difference	$V = \dfrac{W}{Q}$ ✓	Volt, V ✓	$kg\,m^2\,A^{-1}\,s^{-3}$ ✓
Resistance	$R = \dfrac{V}{I}$ ✓	Ohm, Ω ✓	$kg\,m^2\,A^{-2}\,s^{-3}$ ✓

Table 1 Common derived units

Ξ 24/4/16 6/1/17

Worked example

24/4/16

Show that the units of the Planck constant can be written as $kg\,m^2\,s^{-1}$.

Answer

The datasheet gives the Planck constant units as J s and the Planck equation as $E = hf$. Rearranging:

$$h = \frac{E}{f}$$

Inserting the units into the rearranged equation:

6/1/17

$$h = \frac{E}{f} = \frac{J}{s^{-1}}$$

But joules are a derived unit from *work done = force × distance*, as is force (*force = mass × acceleration*), so inserting into the rearranged Planck equation:

$$h = \frac{E}{f} = \frac{J}{s^{-1}} = \frac{kg\,m\,s^{-2}\,m}{s^{-1}} = kg\,m^2\,s^{-1}$$

Prefixes and standard form

Measurements such as the mass of the Earth or the diameter of an atom are either very large or very small compared with the SI base unit, so we use a system of unit prefixes or standard form to express them. The prefixes and their standard form equivalents that you need to know for the examination are shown in Table 2.

Knowledge check 2

What are the base units for (i) acceleration, (ii) power and (iii) potential difference (pd)?

Exam tip

Always use your datasheet to check and confirm the units of constants and derived quantities in questions like this one.

Prefix	Symbol	Standard form equivalent	Example
tera	T	10^{12}	$1\,TW = 10^{12}\,W$
giga	G	10^{9}	$1\,GHz = 10^{9}\,Hz$
mega	M	10^{6}	$1\,M\Omega = 10^{6}\,\Omega$
kilo	k	10^{3}	$1\,km = 10^{3}\,m$
centi	c	10^{-2}	$1\,cm = 10^{-2}\,m$
milli	m	10^{-3}	$1\,mK = 10^{-3}\,K$
micro	μ	10^{-6}	$1\,\mu V = 10^{-6}\,V$
nano	n	10^{-9}	$1\,nm = 10^{-9}\,m$
pico	p	10^{-12}	$1\,pA = 10^{-12}\,A$
femto	f	10^{-15}	$1\,fm = 10^{-15}\,m$

Table 2 Prefixes and standard form equivalents

Conversion factors

Some quantities are expressed in more than one unit. A conversion factor can be used to convert from one set of units to another. Some common conversion factors are shown in Table 3.

Quantity	SI unit	Other common unit	Conversion factor
Energy (quantum scale)	J	eV	$1\,eV = 1.6 \times 10^{-19}\,J$
Energy (domestic)	J	kW h	$1\,kWh = 3.6 \times 10^{6}\,J$
Solar system distances	m	Astronomical unit, AU	$1\,AU = 1.50 \times 10^{11}\,m$
Distance to stars	m	Light year, ly, or parsec, pc	$1\,ly = 9.46 \times 10^{15}\,m$ $1\,pc = 3.26\,ly$

Table 3 Common conversion factors

Summary

- The seven fundamental (base) quantities and their SI units are: mass (kg), length (m), time (s), quantity of matter (mol), temperature (K), electric current (A) and light intensity (cd).
- Derived quantities (such as force) are expressed in terms of the base units $(kg\,m\,s^{-2})$, or they are given a name (newton).
- The SI prefixes and their standard form values are: T (tera, 10^{12}), G (giga, 10^{9}), M (mega, 10^{6}), k (kilo, 10^{3}), c (centi, 10^{-2}), m (milli, 10^{-3}), μ (micro, 10^{-6}), n (nano, 10^{-9}), p (pico, 10^{-12}), f (femto, 10^{-15}).
- Standard unit conversions include joules to electronvolts $(1\,eV = 1.6 \times 10^{-19}\,J)$ and joules to kWh $(1\,kWh = 3.6 \times 10^{6}\,J)$.

Limitation of physical measurements

Random and systematic errors

When an experimental measurement is taken, a measurement error occurs and the measured value is not the 'true value' of the quantity being measured. The 'true 'value' of measurement is the value of the measurement that would be obtained in an ideal world.

There are two types of measurement error:

- **Systematic errors** occur if a measurement is consistently too small or too large. This type of error may be caused by:
  - poor experimental technique
  - zero error on an instrument
  - poor calibration of the instrument.
- **Random errors** occur when repeating the measurement gives an unpredictable and different result. Random errors may arise due to:
  - observer (human) error
  - the readability of the equipment (if a value is changing constantly)
  - external effects on the measured item (e.g. changing temperature).

Systematic errors are reduced by correcting for offsets and using different methods or instruments to obtain the same value, allowing you to compare the results obtained and identifying any systematic error.

Random errors are reduced by taking repeated measurements and averaging.

The quality of measurement

Several different words can be used to assess the quality of a measurement:

- The **accuracy** of a measurement describes how closely the measurement is to the 'true value' of the quantity being measured.
- The **precision** of a measurement describes how closely a number of repeated readings agree with each other. A precise measurement will have very little spread of results around the mean value. However, it does not give any indication of how close to the 'true value' a measurement is. The difference between precision and accuracy is illustrated by the dartboard diagrams in Figure 1.

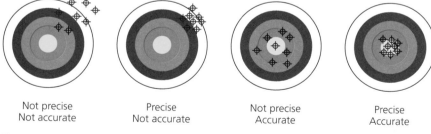

| Not precise | Precise | Not precise | Precise |
| Not accurate | Not accurate | Accurate | Accurate |

Figure 1 The difference between precision and accuracy

- **Repeatability** is the term used when similar measurements are taken by the same person, in the same laboratory over a short period of time, using the same method.
- **Reproducibility** is the term used when similar measurements are made by different people in different laboratories.
- The **resolution** is the smallest change in the quantity being measured by an instrument that gives a measurable change in the value. Using a measuring instrument with a greater resolution will increase the accuracy of the measurement.

Systematic error An error that affects a set of measurements in the same way each time.

Random error An error that affects a measurement in an unpredictable fashion.

Knowledge check 5

Explain the difference between a systematic and a random error.

Accuracy A measurement is accurate if it is considered to be close to the 'true value'.

Precision How close repeated measurements are to each other.

Repeatability The level of consistency of a set of repeated measurements made by the same person, in the same laboratory, using the same method.

Reproducibility The level of consistency of a set of repeated measurements made using the same method by different people in different laboratories.

Resolution The smallest observable change in the quantity being measured by a measuring instrument.

Uncertainty

Error and uncertainty are not the same thing. Error refers to the difference between the measurement of a physical quantity and the 'true value' of that quantity. Uncertainty is a measure of the spread of the value, which is likely to include the 'true value'. A value can be put on the level of uncertainty for a given measurement. For example, if a top-pan balance can measure to a resolution of 0.1 g and records a value of the mass of an object to be 21.6 g, then the measured mass of the copper could be between 21.5 g and 21.7 g. The measurement would be written as: 21.6 g ± 0.1 g.

In this example, the value of uncertainty is an absolute uncertainty. It has the same units as the measurement and represents the range of possible values of the measurement. If a repeated set of measurements is made, then the absolute uncertainty is given as half the range from the highest to the lowest value obtained.

Suppose that a student takes the following measurements of the current flowing through a resistor: 23.1 mA, 23.0 mA, 23.1 mA, 23.2 mA, 23.0 mA.

The mean value of the current is:

$$\frac{23.1 + 23.0 + 23.1 + 23.2 + 23.0}{5} \, mA = 23.08 \, mA$$

However, this value has four significant figures, yet the measurements were taken only to a value of three significant figures, so the mean value must have the same number of significant figures = 23.1 mA.

The absolute uncertainty is therefore

$$\frac{23.2 - 23.0}{2} \, mA = 0.1 \, mA$$

The measurement and absolute uncertainty is therefore 23.1 ± 0.1 mA.

Combining uncertainty

If measurements are being added or subtracted, then to calculate the uncertainty in the overall measurement the absolute uncertainties are added. In other words:

$$(a \pm \Delta a) + (b \pm \Delta b) = (a + b) \pm (\Delta a + \Delta b)$$

$$(a \pm \Delta a) - (b \pm \Delta b) = (a - b) \pm (\Delta a + \Delta b)$$

Where the calculated quantity is derived from the multiplication or division of the measured quantities, the combined percentage error of a calculated quantity is found by adding the percentage uncertainties of the individual measurements.

$$\text{fractional uncertainty} = \frac{\text{absolute uncertainty}}{\text{mean value}}$$

$$\text{percentage uncertainty} = \frac{\text{absolute uncertainty}}{\text{mean value}} \times 100\%$$

Knowledge check 6

The speed of light in a vacuum is a highly reproducible measurement. Explain what this means.

Exam tip

Always give the calculated quantity to the same number of significant figures as the least accurate measured quantity.

Knowledge check 7

Calculate the speed of a sprinter who runs 22.18 m in 2.1 s. Give your answer to an appropriate significant figure.

In general if $a = bc$ or $a = \dfrac{b}{c}$ then

percentage error in a = (percentage error in b) + (percentage error in c)

Worked example

The potential difference across a resistor is measured as $10.0\,V \pm 0.3\,V$. The current through the resistor is measured as $1.3\,A \pm 0.2\,A$. What is the percentage and absolute uncertainty in the power of the resistor?

power = potential difference × current

% uncertainty in pd $= \dfrac{0.3\ V}{10.0\ V} \times 100\% = 3\%$

% uncertainty in current $= \dfrac{0.2\ A}{1.3\ A} \times 100\% = 15\%$

% uncertainty in power = % uncertainty in pd + % uncertainty in current

$= 3\% + 15\%$

$= 18\%$

absolute uncertainty $= \dfrac{\text{value} \times \%\ \text{uncertainty}}{100}$

power $= (13 \pm 2.3)\,W$

Knowledge check 8

A water wave has wavelength of $16.2 \pm 0.1\,m$ and a frequency of $0.35 \pm 0.05\,Hz$. Use $c = f\lambda$ to calculate the speed of the wave, giving its absolute uncertainty.

Uncertainties in graphical data

Uncertainties in data are plotted as error bars on a graph.

General rules for plotting graphs are as follows:

- Choose the axis scales so that plotted points will cover at least half of the graph paper. Using a larger area means that the points can be plotted more accurately.
- Use sensible divisions that can be plotted and read easily (e.g. multiples of 2 or 5).
- Label axes with the plotted quantity and unit in the format 'quantity/unit'.
- Plot points as either a small horizontal cross, +, or a diagonal cross, ×.
- Add error bars to represent the uncertainty — the length of each bar should be the length of the absolute uncertainty for the point.

Estimating uncertainty in gradient

To estimate the uncertainty in the gradient, two additional lines of fit are drawn onto the data points. These are shown in Figure 2.

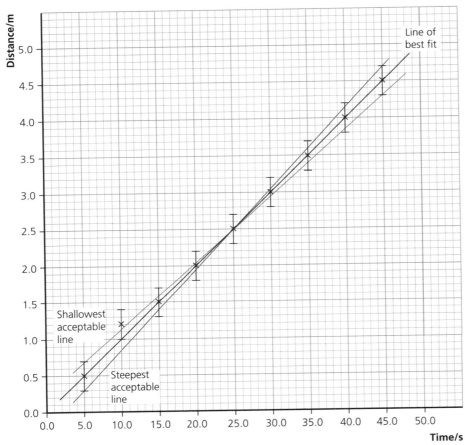

Figure 2 Estimating the uncertainty in the gradient

The points suggest a straight line, so the line of best fit is drawn such that there are an approximately equal number of points on either side of the line.

The point (0,0) is not usually plotted unless it is a point that was measured. If there is a systematic error in the experiment, the line of best fit may not pass through (0,0) and this allows possible systematic errors to be identified.

To calculate the uncertainty in the gradient, two more lines of fit are drawn, one representing the shallowest acceptable line of fit from the bottom of the upper error bar to the top of the lowest error bar (red line in Figure 2) and one representing the steepest acceptable line of fit from the top of the upper error bar to the bottom of the lowest error bar (blue line in Figure 2).

The gradient of both of these lines is calculated and the uncertainty is given by:

$$\text{uncertainty of gradient} = \frac{(\text{maximum gradient} - \text{minimum gradient})}{2}$$

Worked example

A technician is checking the resistance and uncertainty of an unknown fixed resistor using a digital ammeter with a resolution of 0.01 A to measure the current flowing through the resistor and an old analogue voltmeter with a resolution of 0.5 V to measure the potential difference across the resistor. Her data are shown in the table.

Current, I/A; ±0.01 A	0.20	0.40	0.60	0.80	1.00
pd, V/V; ±0.5 V	4.0	6.0	9.0	11.5	14.5

a Plot a graph of pd against current. Use the measurement uncertainties to plot error bars for the points.

b Plot a best fit line and use this to determine a gradient for the line. This is equivalent to the resistance of the fixed resistor.

c Use the best fit line to determine whether the voltmeter had a 'zero error'.

d Use the spread of the error bars to draw two further fit lines, one steeper and one shallower than the best fit line.

e Use the gradients of these fit lines to determine an uncertainty for the resistance of the resistor.

Answer

a A plot of the graph including the error bars (x-axis error bars are almost too small to be seen) is shown Figure 3.

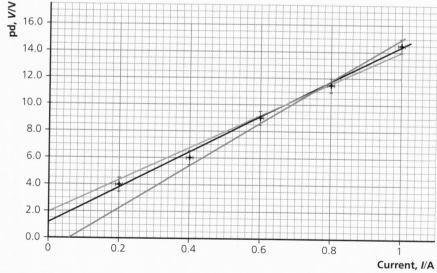

Figure 3 Plotting the graph

➡

b The best fit line gradient (equivalent to the resistance, R) can be determined using measurements from the graph:

$$R = \frac{\Delta V}{\Delta A} = \frac{(14.4 - 1.2)\,\text{V}}{(1.00 - 0.00)\,\text{A}} = 13.2\,\Omega$$

c The best fit line does not go through (0,0) but has a y-axis intercept of 1.2 V, implying that this is the zero error.

d The gradient of the steeper (red) line can be calculated by:

$$R = \frac{\Delta V}{\Delta A} = \frac{(15.0 - 0.0)\,\text{V}}{(1.00 - 0.06)\,\text{A}} = 16.0\,\Omega$$

The gradient of the shallower (blue) line can be calculated by:

$$R = \frac{\Delta V}{\Delta A} = \frac{(14.0 - 2.0)\,\text{V}}{(1.00 - 0.00)\,\text{A}} = 12.0\,\Omega$$

e The uncertainty of the resistance can be determined using the two fit line gradients:

$$\text{uncertainty of gradient} = \frac{(16.0 - 12.0)}{2} = 2.0\,\Omega$$

This implies that the value of the resistance of the resistor is $(13.2 \pm 2.0)\,\Omega$.

Summary

- A systematic error affects a set of measurements in the same way each time. A random error affects a measurement in an unpredictable fashion.
- A measurement is accurate if it is considered to be close to the 'true value'.
- The resolution of a measuring instrument is the smallest observable change in the quantity that changes the reading on an instrument.
- Uncertainty is a measure of the spread of the value which is likely to include the 'true value'.

- The absolute uncertainty of a measurement represents the range of possible values of the measurement.
- The fractional and percentage uncertainties of a measurement are defined as:

$$\text{fractional uncertainty} = \frac{\text{absolute uncertainty}}{\text{mean value}}$$

$$\text{percentage uncertainty} = \frac{\text{absolute uncertainty}}{\text{mean value}} \times 100\%$$

- Uncertainty in a data point on a graph can be shown using error bars.

Estimation of physical quantities

Orders of magnitude

If a quantity is an **order of magnitude** bigger than another quantity, then it is about ten times bigger. For example, the diameter of the Sun is an order of magnitude bigger than the diameter of Saturn. Two orders of magnitude difference would be 100 times larger, or 10^2, and three orders of magnitude would be 1000 times bigger, or 10^3 etc.

Estimating physical quantities

You need to be able to estimate the approximate value of quantities. You can do this easily by comparing the value that you are estimating to values that you know well. Table 4 gives some common quantities to base your estimates on.

Length	Mass	Temperature
Width of a human hair $\approx 0.1\,mm$	Mass of an apple $\approx 100\,g$	Freezer temperature $\approx -20°C$
Height of a door handle $\approx 1\,m$	Mass of a bag of sugar $\approx 1\,kg$	Room temperature $\approx 20°C$
Length of a football pitch $\approx 100\,m$	Mass of a small car $\approx 1000\,kg$	Candle flame temperature $\approx 1000°C$

Table 4

Approximated estimated values are usually given as 'orders of magnitude' – in other words, to the nearest power of 10. For example, a three-storey town house would have each storey approximately three 'door-handles' high, so the height of the house would be about 10 m.

Derived quantities can also be estimated. For example, you could estimate how much work you do against gravity climbing a ladder to the top of a three-storey town house. The gravitational potential energy at the top of the house is given by $\Delta E_p = mg\Delta h$. We have already estimated Δh to be 10 m, and an adult human has a mass of approximately 50 kg. As $g \approx 10\,m\,s^{-2}$, $\Delta E_p = mg\Delta h = 50\,kg \times 10\,m\,s^{-2} \times 10\,m = 5000\,J$.

Summary

- An order of magnitude difference in a quantity involves one quantity being ten times bigger than the other.
- The approximate values of physical quantities can be estimated using common everyday values that are known.

One quantity is one **order of magnitude** bigger than another if it is about ten times bigger.

Knowledge check 9

A nearby star is 5 ly away. A star towards the edge of the Milky Way is 50 000 ly away. What is the order of magnitude difference between these two distances?

Exam tip

When you are estimating a quantity, use everyday values that you know well to base your estimates on.

■ Particles and radiation

In **Particles and radiation** you will study the fundamental structure and behaviour of matter and radiation on the atomic and nuclear scale. You need to know how the structure of matter is dictated by the combinations of the fundamental particles, such as quarks and electrons, which form the basic building blocks of matter and the forces that interact between them. In particular, you need to know about the electromagnetic interaction between particles, which is governed by the quantum nature of photons of electromagnetic radiation.

Particles

Constituents of the atom

The simplest model of the atom was proposed by the Greek philosopher Democritus about 2 500 years ago. His model was the first to use the word 'atomos' and it viewed matter as made up of tiny, indivisible spheres. The classic model of the atom used extensively in modern culture today is the Rutherford–Bohr model (Figure 4), first proposed by Niels Bohr in 1913.

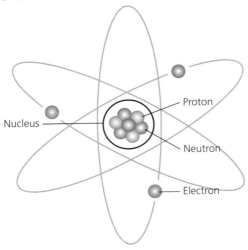

Figure 4 Rutherford–Bohr model of the atom

In this atomic model, a tiny nucleus is surrounded by electrons arranged in specific orbits around the nucleus. The model was based on experiments performed by Ernest Rutherford, Hans Geiger and Ernest Marsden two years earlier in 1911, in which they fired alpha particles at thin films of gold. The alpha particles probed the structure of the gold film by scattering as they passed through the film. Rutherford was able to analyse the scattering patterns and proposed that atoms of gold have a radius of about 135×10^{-12} m (135 pm) but that most of the atom was in fact empty space, and that the mass of the atom is concentrated into a tiny nucleus (the gold nuclear radius is 27×10^{-15} m or 27 fm), about 5000 times smaller than the corresponding atom. Rutherford also proposed that the nucleus was made up of protons as it had a positive charge.

Niels Bohr refined this model by adding the orbiting structure of the electrons. In 1932, James Chadwick completed this simple model with his discovery of the neutron. Protons and neutrons are particles found in the nucleus and are given the collective name **nucleons**.

Protons, neutrons and electrons

Key physical data for protons, neutrons and electrons are given in Table 5.

Sub-atomic particle	Charge (C)	Relative charge	Mass (kg)
Proton	$+1.60 \times 10^{-19}$	$+1$	1.673×10^{-27}
Neutron	0	0	1.675×10^{-27}
Electron	-1.60×10^{-19}	-1	9.11×10^{-31}

Table 5

Specific charge

The specific charge of a particle is defined as the charge per unit mass and its units are $C\,kg^{-1}$. Specific charge is calculated using the formula:

$$\text{specific charge of a particle} = \frac{\text{charge of the particle}}{\text{mass of the particle}} = \frac{Q}{m}$$

So the specific charge of an electron is calculated by:

$$\text{specific charge of an electron} = \frac{-1.60 \times 10^{-19}\,C}{9.11 \times 10^{-31}\,kg} = -1.76 \times 10^{11}\,C\,kg^{-1}$$

The specific charge of the neutron is 0 and the specific charge of the proton is $+9.56 \times 10^{7}\,C\,kg^{-1}$.

Worked example

A physicist is measuring the specific charge of particles emitted during radioactive decay.

a State what is meant by the term 'specific charge of a particle' and give an appropriate unit for specific charge.

b Calculate the specific charge of the carbon-14 nucleus.

c Explain in terms of proton and neutron why the specific charge of the nucleus changes during the beta minus decay of carbon-14 into a nitrogen isotope.

d Deduce the number of protons and neutrons in the nitrogen isotope.

Answer

a The specific charge of a particle is its ratio of charge to mass and its unit is $C\,kg^{-1}$.

b Carbon-14 contains six protons and eight neutrons, so:

$$\text{specific charge} = \frac{\left(6 \times 1.6 \times 10^{-19}\,C\right)}{\left(6 \times 1.673 \times 10^{-27}\,kg\right) + \left(8 \times 1.675 \times 10^{-27}\,kg\right)} = 4.1 \times 10^{7}\,C\,kg^{-1}$$

A **nucleon** is a particle that exists in the nucleus.

Knowledge check 10

Which two particles are nucleons?

Exam tip

When doing calculations involving numbers with different significant figures, always state your final answer to the same number of significant figures as the data with the least number of significant figures.

c During beta minus decay a neutron decays into a proton and emits a beta minus particle and an electron antineutrino. This means that the overall mass of the nucleus goes down very slightly (by 0.002×10^{-27} kg), but the overall charge increases by +1 (1.6×10^{-19} C). The overall effect of this change is to increase the specific charge.

d The nitrogen isotope contains seven protons and seven neutrons.

Exam tip

In some specific charge calculations, the combined mass of the electrons is insignificant compared with the mass of the protons and neutrons, and can effectively be ignored. For particles with very high numbers of electrons, the mass is not insignificant and must be taken into account.

Notation for nuclei and isotopes

The number of protons in any given nucleus is called the **proton number**, Z. The total number of protons and neutrons in the nucleus is called the **nucleon number**, A. These two numbers identify any nucleus uniquely. The number of neutrons in a nucleus can be calculated by subtracting the proton number from the nucleon number. The two nuclear numbers and the chemical symbol of the element are used together as a shorthand way of describing any nucleus, called the $^{A}_{Z}X$ notation. The nucleus of a given element, with a specific value of A and Z, is called a **nuclide**.

The $^{A}_{Z}X$ notation also allows us to distinguish between **isotopes**. Isotopes are nuclei (and atoms) of the same element having the same proton number, Z (and therefore chemical symbol, X), but different numbers of neutrons and hence different nucleon numbers, A.

Using isotopic data

Although there are nine different possible isotopes of the metallic element lithium (from $^{4}_{3}Li$ to $^{12}_{3}Li$), only two occur naturally (Table 6).

Isotope of lithium	Number of neutrons	Relative abundance
$^{6}_{3}Li$	3	7.5%
$^{7}_{3}Li$	4	92.5%

Table 6

All the isotopes have the same number of protons ($Z = 3$), but they have different numbers of neutrons.

Knowledge check 11

State the equation for calculating the specific charge of a particle and use it to calculate the specific charge of a deuteron ($^{2}_{1}H$ nucleus).

The **proton number** of a nucleus is the number of protons in the nucleus.

The **nucleon number** of a nucleus is the number of protons plus the number of neutrons.

A **nuclide** is the nucleus of a given element with specified values of A and Z.

Isotopes are nuclei or atoms with the same number of protons, but different numbers of neutrons.

Knowledge check 12

What is an isotope?

Knowledge check 13

Use the $^{A}_{Z}X$ notation to describe the three isotopes of hydrogen: hydrogen-1, deuterium (hydrogen-2) and tritium (hydrogen-3).

Summary

- An atom is made up of a nucleus containing protons and neutrons, surrounded by electrons, and is represented using the $_Z^A X$ notation.
- The proton number, Z, of a nuclide is the number of protons inside the nuclide. The nucleon number, A, of a nuclide is the number of protons plus the number of neutrons.
- The specific charge of a particle is the total (overall) charge on the particle divided by the total mass of the particle.
- Isotopes are atoms of the same element, containing the same number of protons (and electrons) but different numbers of neutrons.

Stable and unstable nuclei

The strong nuclear force

The strong nuclear force (strong force) is the fundamental force of nature that holds nucleons together. It is a force that exists only between nucleons (and the fundamental particles that make up nucleons, called quarks). The force is extremely short range – up to about 3 fm (1 fm, or 1 femtometre, = 1×10^{-15} m). The force is attractive up to separations of about 0.5 fm, holding the nucleons together inside the nucleus, and is then repulsive for separations less than 0.5 fm, keeping the nucleons (and the nucleus) from imploding in on themselves (Figure 5).

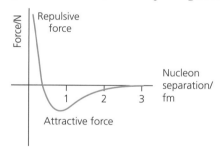

Figure 5 Graph of the strong nuclear force

Unstable nuclei

Many nuclei are unstable. This means that the nucleus will change its nucleon composition and emit nuclear radiation in the form of alpha (α), beta (β), or gamma (γ) radiation. The emission of the nuclear radiation makes the nucleus more stable, although some nuclei, particularly with high proton numbers, can undergo a series of decays, progressively producing daughter nuclei that are more and more stable — this is called a decay series.

During alpha decay, the nucleus ejects two protons and two neutrons joined together. This particle is called an alpha particle and it has the same structure as a helium nucleus. The $_Z^A X$ notation for an alpha particle is $_2^4 He$ and the general equation of nuclear decay via alpha emission is:

$$_Z^A X \rightarrow {}_{Z-2}^{A-4} Y + {}_2^4 He$$

You can see that the nucleon number A decreases by 4 and the proton number decreases by 2.

Knowledge check 14

What is the strong nuclear force?

Knowledge check 15

Describe how a nucleus of radium-226 decays via alpha emission into radon-222.

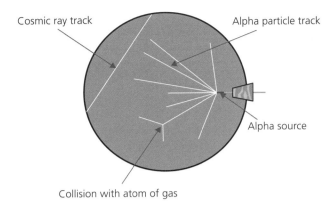

Cosmic ray track

Alpha particle track

Alpha source

Collision with atom of gas

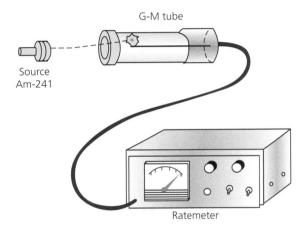

G-M tube

Source
Am-241

Ratemeter

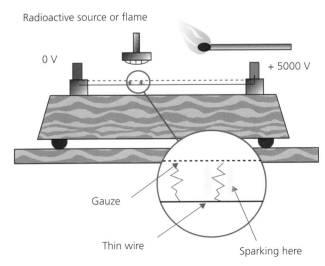

Radioactive source or flame

0 V

+ 5000 V

Gauze

Thin wire

Sparking here

Figure 6 A cloud chamber, Geiger counter and spark counter detecting the range of alpha particles

Alpha particles have a very short range in air, which can be shown experimentally using a cloud chamber, a Geiger counter or a spark counter (Figure 6).

Many more higher proton number nuclei decay via alpha emission than low proton number nuclei. These are often beta minus (β^-) emitters. Beta minus emission involves a neutron within the nucleus decaying into a proton and emitting an electron and a second particle called an antineutrino, $\bar{\nu}$.

The $^{A}_{Z}X$ notation for a beta minus particle is $^{0}_{-1}e$, and the general equation of nuclear decay via beta minus emission is:

$$^{A}_{Z}X \rightarrow {}^{A}_{Z+1}Y + {}^{0}_{-1}e + \bar{\nu}_e$$

You can see that the nucleon number A stays constant and the proton number increases by 1.

Neutrinos and their antiparticles are neutral, almost massless, fundamental particles that rarely interact with other forms of matter. They were first proposed by Wolfgang Pauli in 1930 to address the problem of the apparent violation of the law of conservation of energy. As two particles are emitted, they share the available energy, as shown in Figure 7.

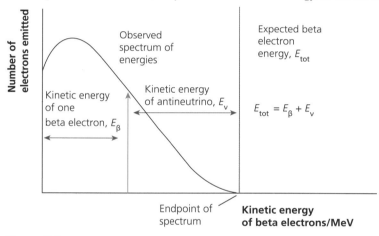

Figure 7 The beta particle emission spectrum

Until Pauli's proposal of the neutrino, beta particles were observed to be emitted with a range of different energies, up to a maximum energy. Those beta particles emitted at lower than maximum energy were thought to violate the law of conservation of energy, but in reality the extra energy is taken up by the neutrinos.

Worked example

Neutral atoms of the radioactive gas radon-222 can be represented by $^{222}_{86}Rn$.

a State the types and number of constituent particles in an atom of radon-222.

b Radon-222 decays via alpha emission forming an isotope of polonium, Po. Write down an equation that represents this decay.

c $^{A}_{Z}Rn$ is another isotope of radon. State a possible value for A and Z.

Answer

a Z is the number of protons, hence 86 protons. The atom is neutral so there must be 86 electrons. A is the nucleon number so there are $(222 - 86) = 136$ neutrons.

$\rightarrow$

b $^{222}_{86}\text{Rn} \rightarrow \,^{218}_{84}\text{Po} + \,^{4}_{2}\text{He}$

c As the isotope is still radon, $Z = 86$. Radon has isotopes from $A = 195$ up to $A = 229$. It is usually best to pick a value ±1, representing one more or one less neutron.

Summary

- Nuclei are held together by the strong nuclear force, but some nuclei are unstable and can decay via radioactive decay.
- Alpha decay involves the emission of a helium nucleus from a nuclide; beta minus decay involves the decay of a neutron into a proton and the emission of an electron and an electron antineutrino.
- The existence of neutrinos was proposed to account for the conservation of energy in beta decay.

Particles, antiparticles and photons

All particles have a corresponding antiparticle. Antiparticles have the same mass and rest-energy as their corresponding particle, but they have the opposite charge (and opposite lepton and baryon numbers). The antiparticle of the electron, the positron, is the only antiparticle to have its own name — all the other antiparticles are called the anti version of their particle, for example, the antiproton and the antineutrino (Table 7).

Particle	Electron, e⁻	Proton, p	Neutron, n	Neutrino, v
Antiparticle	Positron, e⁺	Antiproton, $\bar{p}$	Antineutron, $\bar{n}$	Antineutrino, $\bar{v}$

Table 7

The rest-energy of a particle is the equivalent energy that the particle would have if all of its mass was converted into energy. Rest-mass energy is usually measured in MeV (mega electronvolts). 1 MeV is equal to 1×10^{-13} J. The rest-energy of the electron and the proton (and their antiparticles) are 0.51 MeV and 938 MeV respectively.

Photons, annihilation and pair creation

On the nuclear scale, electromagnetic radiation can also behave like particles. The 'particles' of electromagnetic radiation are called photons. The energy of a photon is related to its frequency, f, or its wavelength, λ, by the equation:

$$E = hf = \frac{hc}{\lambda}$$

where c is the speed of light and h is the Planck constant ($h = 6.63 \times 10^{-34}$ J s).

Gamma ray photons can be detected using a Geiger counter or a scintillation counter.

When a particle meets its antiparticle they will **annihilate** each other and their mass will be converted into energy, in the form of gamma ray photons. During annihilation two photons will always be produced in order that the law of conservation of momentum is obeyed. The two photons will have equal energies and if the particle and its antiparticle are at rest when they annihilate each other, the two photons will be emitted in opposite directions.

Worked example

The table below gives information about some particles.

Particle	Fundamental?	Relative charge
Proton	No	+1
Neutron	No	0
Electron	Yes	−1

a Give the name of the non-nuclear particle and its antiparticle.

b Give one property of the antiparticle that is the same for its corresponding particle and one property that is different.

c Write an equation showing the annihilation of the non-nuclear particle and its antiparticle.

Answers

a The non-nuclear particle is the electron; its antiparticle is the positron.

b The electron and the positron have the same mass but opposite charges (electron, −1; positron, +1).

c $_{-1}^{0}e + _{+1}^{0}e \rightarrow 2\gamma$

PET (positron emission tomography) scanners, often found in hospital imaging departments, use the annihilation of electrons and positrons within the body to image internal body structures. A positron-emitting radionuclide is attached to a biologically active molecule such as glucose. The active molecule accumulates round a target organ or damaged site, emits positrons that immediately meet electrons and annihilate. The detecting scanners around the outside of the machine detect the two gamma rays, emitted in opposite directions, and the emission site can be pinpointed using computer software. The detectors rotate around the body and create a map of the emission site, as shown in Figure 8.

Knowledge check 18

How does the PET scanner pinpoint the site of the electron–positron emission?

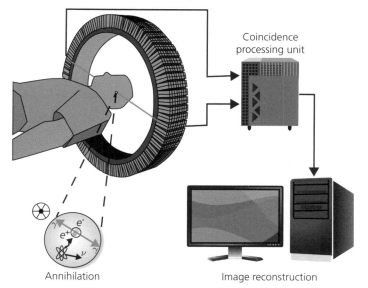

Annihilation   Image reconstruction

Figure 8 PET scanner

Under certain circumstances, the opposite of annihilation — **pair creation** — can occur. During pair creation a high-energy gamma ray photon can interact with a large stationary nucleus (which acts in a way similar to a chemical catalyst, in that it provides a site for the pair creation to occur) and it can convert itself into a particle and its antiparticle pair.

Summary

- For every type of particle there is a corresponding antiparticle.
- Particle–antiparticle pairs can annihilate each other, producing photons of electromagnetic radiation, and can be produced from photons via pair creation.
- Photons of electromagnetic radiation have an energy, $E = hf = \dfrac{hc}{\lambda}$.

Particle interactions

Fundamental interactions

There are four fundamental ways that particles of matter can interact with each other. Each of the interactions occurs due to the transfer of 'exchange' particles. These are summarised in Table 8.

Interaction	Responsible for	Exchange particle
Electromagnetic	The force between charged particles	Virtual photon
Strong nuclear (known as the 'strong' interaction)	The force between quarks and nucleons	Gluon
Weak nuclear (known as the 'weak' interaction)	Radioactive decay and nuclear fusion	W^-, W^+ or Z^0 particles
Gravity	Gravitational force between masses	Graviton

Table 8

Exam tip

You will not be asked about Z^0 and gravitons in the examination.

Exchange particles are transferred forwards and backwards between the interacting particles and the size of the force is dictated by the rate of exchange. Exchange particles are sometimes called virtual particles because they exist only during the interaction. They are emitted by one of the interacting particles and are absorbed by the other interacting particle.

Feynman diagrams

The interactions between particles can be shown diagrammatically using **Feynman diagrams**, named after the American physicist Richard Feynman. The decay of a neutron during beta minus decay is shown in the Feynman diagram in Figure 9.

Figure 9 A Feynman diagram showing beta minus decay

During beta minus decay a neutron decays into a proton and a W⁻ exchange particle. The W⁻ is unstable and subsequently decays into an electron, e⁻, and an anti-electron neutrino, $\overline{v}_e$. Figure 9 summarises the equation shown below:

$$^1_0n \rightarrow {}^1_1p + {}^{\;0}_{-1}e + \overline{v}_e$$ anti

Feynman diagrams show the path of particles as straight lines with arrows (showing the sequence of interaction) and exchange particles as wavy lines. All the lines are labelled with the particle. In Feynman diagrams time generally flows from below upwards or from left to right. The length of the lines has no meaning, but the intersections between the lines do — this is where particles are annihilated or created.

The weak interaction

There are four main weak interactions that you need to know. The first is the decay of a neutron during beta minus decay shown above. The other three weak interactions are illustrated by the Feynman diagrams shown in Figures 10, 11 and 12.

Figure 10 A Feynman diagram showing beta plus decay

In the case shown in Figure 10, a proton decays into a neutron and a W⁺ exchange particle, which subsequently decays into a positron and an electron neutrino. This is summarised by the equation:

$$^1_1p \rightarrow {}^1_0n + {}^{\;0}_{+1}e + v_e$$

During electron capture, an electron is absorbed by a proton within a nucleus. The proton decays into a neutron and a W⁺ exchange particle, which interacts with the electron forming an electron neutrino:

$$^1_1p + {}^{\;0}_{-1}e \rightarrow {}^1_0n + v_e$$

This is shown by the Feynman diagram in Figure 11.

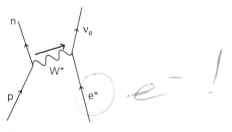

Figure 11 A Feynman diagram showing electron capture

A **Feynman diagram** is a schematic way of illustrating a particle interaction. Particles are represented by straight lines with arrows on and exchange particles as wavy lines. Annihilation and pair creation occur at the vertices.

Exam tip

β⁻ decay is negative so involves e⁻, an antineutrino and W⁻, whereas β⁺ decay is positive so involves e⁺, a neutrino and W⁺

Knowledge check 19

Explain the difference between beta minus emission and beta plus emission.

Content Guidance

The last weak interaction that you need to know is electron–proton collision, during which an electron and a proton collide, transferring a W^- exchange particle. The proton then decays into a neutron and the electron decays into an electron neutrino, as shown by the equation and the Feynman diagram in Figure 12.

$${}^1_1p + {}^0_{-1}e \rightarrow {}^1_0n + \nu_e$$

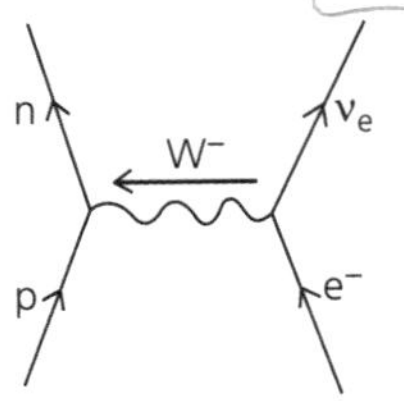

Figure 12 A Feynman diagram showing an electron–proton collision

Worked example

Exchange particles can be used to explain the interactions between particles.

a Complete the following table by identifying the particle interaction from the exchange particle.

Exchange particle	Interaction
W$^+$	
Photon	

b Identify the values A and Z in the following equation summarising the beta plus decay of magnesium-23:

$${}^{23}_{12}Mg \rightarrow {}^A_Z Na + {}^0_{+1}e + \nu_e$$

c During the decay shown above, a proton decays into a neutron and a W^+ exchange particle. Draw a Feynman diagram to illustrate this decay.

Answer

a

Exchange particle	Interaction
W+	Weak
Photon	Electromagnetic

b (Conservation of) $A = 23$. (Conservation of Z): $12 = Z + 1 \Rightarrow Z = 11$.

c A proton decays into a neutron and a W^+ exchange particle, which subsequently decays into a positron and an electron neutrino (Figure 13).

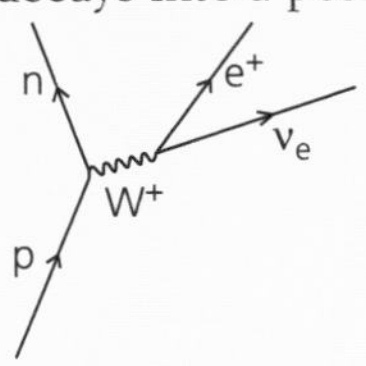

Figure 13 Feynman diagram illustrating beta-plus decay

Summary

- Particles of matter can interact through the four fundamental interactions: gravity, electromagnetic, weak nuclear and strong nuclear. Each interaction occurs due to the transfer of exchange particles.
- The exchange particle of the electromagnetic force is the photon and the exchange particles for the weak interaction are the W^+ and W^- particles.
- Weak interactions include beta minus, beta plus, electron capture and electron–proton collisions.
- Particle interactions can be represented by Feynman diagrams.

Classification of particles

The Particle Garden

During the second half of the twentieth century, scientists and engineers around the world worked in collaboration to identify a large range of new sub-nuclear particles. The particles eventually began to be identified as belonging to a number of groups, all with similar properties. The 'Particle Garden' illustrating these groups of particles is shown in Figure 14.

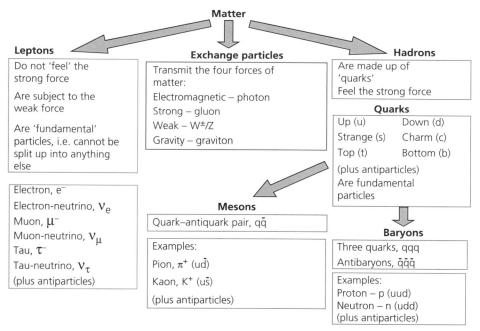

Figure 14 The Particle Garden

Hadrons

Hadrons are particles that are subject to strong interaction. There are two classes of hadrons:

- **baryons** (such as the proton, neutron) and their antibaryons (the antiproton and antineutron)
- **mesons** (such as the pion and the kaon and their antiparticles).

A **baryon** is a particle made up of three quarks.

A **meson** is a particle made up of a quark–antiquark pair.

The proton is the only stable baryon into which other baryons eventually decay, The mean lifetime of most other baryons (with the exception of the neutron) is very short, of the order of 10^{-12} s. The neutron has a mean lifetime of 880 s.

All baryons have a quantum number associated with them called the **baryon number, B** — this distinguishes them as baryons. All baryons have a baryon number of +1, all antibaryons have a baryon number of −1. All non-baryons have a baryon number of 0. During all particle interactions the total baryon number of the initial particles must equal the total baryon number of the final particles. This is called the conservation of baryon number.

The pion meson and its antiparticle, the antipion, are particles produced by protons and neutrons within nuclei and act as the exchange particle of the strong nuclear force between nucleons. The kaon meson is a strange particle (see p. 29) found commonly in cosmic rays from space in the upper atmosphere. They decay into pions.

(see p. 29)

Worked example

Quarks are fundamental particles that make up hadrons, which can be either baryons or mesons.
a What is the quark structure of a baryon?
b What is the quark structure of a meson?
c What property defines hadrons?
d State the quark structure of:
   a neutron
   a proton
e Give two properties of hadrons that distinguish them from leptons.

Answers

a Baryons are made up of three quarks, qqq.
b Mesons are made up of a quark–antiquark pair, $q\bar{q}$.
c Hadrons are subject to the strong interaction.
d Neutron = udd; proton = uud.
e Leptons are not made up of quarks; leptons do not experience the strong force. Hadrons are not fundamental particles; all hadrons (eventually) decay into protons.

Leptons

Leptons are particles that are subject to the weak interaction. They consist of:
- electron, e^-
- electron–neutrino, v_e
- muon, μ^-
- muon–neutrino, v_μ
- tau, τ^-
- tau–neutrino, v_τ

plus all their antiparticles.

Like baryons, leptons have an associated quantum number that distinguishes them as leptons. All leptons have a lepton number, L, of +1, antileptons, −1, and all non-leptons, 0. The lepton number is always conserved in particle interactions, so the total lepton number before an interaction is the same as the total lepton number after an interaction.

Knowledge check 20

Explain the difference between baryons and mesons.

Knowledge check 21

What is the baryon number of a pion?

Exam tip

You will not be asked any questions in the examination on the tau lepton or the tau–neutrino lepton.

Knowledge check 22

What is the difference between a hadron and a lepton?

Muon leptons are particles that decay into electrons (electrons are thought to be stable).

Strange particles

Strange particles are particles that are produced through the strong interaction but decay through the weak interaction. Kaons are one of the more common strange particles as they are produced by cosmic rays in the upper atmosphere. Strange particles are given a **strangeness** quantum number (symbol S), to reflect the fact that strange particles are always created in pairs. Strangeness is always conserved during strong interactions (the total strangeness before an interaction = the total strangeness after an interaction), but can change by 0, +1 or −1 during weak interactions.

Cosmic rays

As cosmic rays are high-energy particles coming from space, consisting mostly of protons and very high-energy light atomic nuclei, they interact with matter in the upper atmosphere, producing showers of other particles. The paths of cosmic rays can be captured as tracks in a cloud chamber.

A schematic diagram of a cosmic ray shower, showing how pions (and kaons) can be produced, is shown in Figure 15.

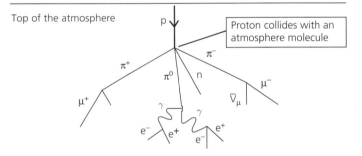

Figure 15 A cosmic ray shower in the upper atmosphere

More modern versions of cloud chambers can be mounted into high-altitude aircraft and balloons and these can then be used to study the interactions of cosmic rays with the upper atmosphere.

Strangeness is a property of certain particles that is used to describe their formation via the strong interaction but their decay via the weak interaction.

Exam tip

In the examination paper you will be provided with any data such as charge, Q, baryon number, B, lepton number, L, and strangeness, S, of individual particles.

Summary

- Particles are classified as hadrons, leptons or exchange particles.
- Hadrons are made up of quarks. Three quarks make up a baryon and a quark–antiquark pair makes up a meson. Hadrons are subject to the strong interaction.
- Like charge Q, baryon number B is conserved in particle interactions.
- The proton is the only stable baryon into which other baryons eventually decay.

- Electrons, muons, tau and neutrinos are all leptons. Lepton number L is also conserved in particle interactions. Leptons are subject to the weak interaction.
- Muons are particles that decay into electrons.
- Strange particles, like kaons, are formed via the strong interaction but decay via the weak interaction.
- Strangeness, S, is conserved in strong interactions but can change by 0, +1 or −1 in weak interactions.

Quarks and antiquarks

Quarks are fundamental particles that make up hadrons, such as protons and neutrons. There are six quarks (and their six antiquarks):

- up, u
- down, d
- strange, s
- charmed, c
- bottom, b
- top, t

Baryons are made up of three quarks (qqq) and antibaryons are made up of three antiquarks ($\bar{q}\bar{q}\bar{q}$). Mesons are made up of a quark–antiquark pair ($q\bar{q}$).

The charge Q, baryon number B, and strangeness S for the up, down and strange quarks and their antiquarks are shown in Table 9.

Quark	Charge, Q	Baryon number, B	Strangeness, S
Up, u	$+\frac{2}{3}$	$+\frac{1}{3}$	0
Anti-up, $\bar{u}$	$-\frac{2}{3}$	$-\frac{1}{3}$	0
Down, d	$-\frac{1}{3}$	$+\frac{1}{3}$	0
Anti-down, $\bar{d}$	$+\frac{1}{3}$	$-\frac{1}{3}$	0
Strange, s	$-\frac{1}{3}$	$+\frac{1}{3}$	-1
Anti-strange, $\bar{s}$	$+\frac{1}{3}$	$-\frac{1}{3}$	$+1$

Table 9

The combinations of quarks for the hadrons that you need to know for the exam are shown in Table 10.

Hadron	Quark combination	Charge, Q	Baryon number, B	Strangeness, S
Proton	uud	$+\frac{2}{3}+\frac{2}{3}-\frac{1}{3}=+1$	$+\frac{1}{3}+\frac{1}{3}+\frac{1}{3}=+1$	$0+0+0=0$
Antiproton	$\bar{u}\bar{u}d$	$-\frac{2}{3}-\frac{2}{3}+\frac{1}{3}=-1$	$-\frac{1}{3}-\frac{1}{3}-\frac{1}{3}=-1$	$0+0+0=0$
Neutron	udd	$+\frac{2}{3}-\frac{1}{3}-\frac{1}{3}=0$	$+\frac{1}{3}+\frac{1}{3}+\frac{1}{3}=+1$	$0+0+0=0$
Antineutron	$\bar{u}\bar{d}\bar{d}$	$-\frac{2}{3}+\frac{1}{3}+\frac{1}{3}=0$	$-\frac{1}{3}-\frac{1}{3}-\frac{1}{3}=-1$	$0+0+0=0$
Pion plus	$u\bar{d}$	$+\frac{2}{3}+\frac{1}{3}=+1$	$+\frac{1}{3}-\frac{1}{3}=0$	$0+0+0=0$
Pion minus	$d\bar{u}$	$-\frac{1}{3}-\frac{2}{3}=-1$	$+\frac{1}{3}-\frac{1}{3}=0$	$0+0+0=0$
Kaon plus	$u\bar{s}$	$+\frac{2}{3}+\frac{1}{3}=+1$	$+\frac{1}{3}-\frac{1}{3}=0$	$0+1=+1$
Kaon minus	$s\bar{u}$	$-\frac{1}{3}-\frac{2}{3}=-1$	$+\frac{1}{3}-\frac{1}{3}=0$	$-1+0=-1$

Table 10

Exam tip

You will only be asked questions on the up, down and strange quarks.

Knowledge check 23

Use the quark model to explain why protons have a relative charge of +1.

Beta decay and quarks

During beta minus decay a neutron decays into a proton and a W^- exchange particle that then subsequently decays into an electron and an electron antineutrino, which is summarised by the equation:

$$^1_0n \rightarrow \,^1_1p + \,^0_{-1}e + \overline{\nu}_e$$

Studying the quark structure of the neutron (udd) and the proton (uud) shows us that inside the neutron, a down quark, d, decays into an up quark, u, and the W^- particle. This means that the decay equation could be written as:

$$d \rightarrow u + e^- + \overline{\nu}_e$$

and can be summarised by the Feynman diagram in Figure 16.

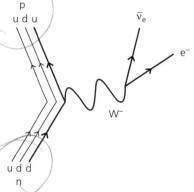

Figure 16 Feynman diagram showing beta minus decay in terms of quarks

Beta plus decay can be illustrated in a similar way — a proton decays into a neutron and a W^+ exchange particle, which then decays into a positron and an electron neutrino (see Figure 17):

$$^1_1p \rightarrow \,^1_0n + \,^0_{+1}e + \nu_e$$

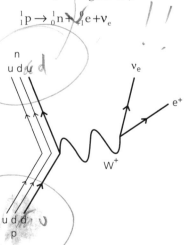

Figure 17 Feynman diagram showing beta plus decay in terms of quarks

Knowledge check 24

Describe the change in quarks that occurs during beta plus decay.

Content Guidance

Worked example

The kaon plus, K⁺, is a meson with strangeness +1.

a State the quark structure of the kaon plus meson. *u s̄*

    b/d̄ s̄

b State the charge Q, baryon number B, and the strangeness number S, of a kaon plus meson. *+1, 0, +1*

c The Feynman diagram for the decay of the anti-strange quark in a kaon plus meson is shown in Figure 18. State the name of the interaction responsible for this decay. *weak*

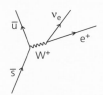

Figure 18 Feynman diagram for the decay of the anti-strange quark in a kaon plus meson

d Energy and momentum are conserved when the W⁺ particle is produced. State two other quantities that are also conserved and one that is not. Write an equation for this decay. *Charge, Baryon, Lepton, Strangeness*

    s̄ → n̄ + νe + e⁺

    −1

Answers

a The kaon plus meson has the quark structure u s̄.

b The u quark has the properties $Q = +\frac{2}{3}$, $B = +\frac{1}{3}$, $S = 0$. The s̄ quark has the properties $Q = +\frac{1}{3}$, $B = -\frac{1}{3}$, $S = +1$.

c The W+ exchange particle is responsible for the weak interaction.

d Charge is always conserved; baryon number is conserved; strangeness is not conserved.

Summary

- Quarks (and antiquarks) are particles that make up hadrons.
- The up (u), down (d) and strange (s) quarks have the following properties (their antiparticles have opposite values to the quarks):

Quark	Charge, Q	Baryon number, B	Strangeness, S
Up (u)	$+\frac{2}{3}$	$+\frac{1}{3}$	0
Down (d)	$-\frac{1}{3}$	$+\frac{1}{3}$	0
Strange (s)	$-\frac{1}{3}$	$+\frac{1}{3}$	−1

- The quark structure of a proton is uud; the neutron is udd; the antiproton is ūūd̄; the antineutron is ūd̄d̄; the pion plus is ud̄; and the kaon plus is us̄.
- During beta minus decay, a neutron decays into a proton and a W⁻ particle, which subsequently decays into an electron antineutrino and an electron.

Applications of conservation laws

During particle interactions various particle quantities are conserved. In all particle interactions the following quantities are always conserved:

- mass–energy, m/E
- momentum, p
- charge, Q
- baryon number, B
- lepton number, L

During strong and electromagnetic interactions, strangeness S is conserved, but during weak interactions it can be conserved, or it can change by $+1$ or -1. You would be given any particle quantities that you would need for the examination either as part of the question or on your datasheet. The conservation laws can be used to determine whether a particle interaction can occur or not.

Worked example

A student is checking to see which of the following equations are correct for beta plus decay:

$A: {}_1^1p \rightarrow {}_0^1n + {}_{+1}^0e + \overline{\nu}_e$

$B: {}_1^1p \rightarrow {}_0^1n + {}_{+1}^0e + \nu_e$

Use the particle conservation laws to determine the correct equation.

Answers

Use a conservation table to check this:

$A: {}_1^1p \rightarrow {}_0^1n + {}_{+1}^0e + \overline{\nu}_e$

Quantity	Before interaction $_1^1p$	Total before interaction	After interaction $_0^1n$	$_{+1}^0e$	$\overline{\nu}_e$	Total after interaction	Conserved?
Q	+1	+1	0	+1	0	+1	✓
B	+1	+1	+1	0	0	+1	✓
L	0	0	0	−1	−1	−2	✗
S	0	0	0	0	0	0	✓

Lepton number is not conserved, so the interaction is not permitted to occur.

$B: {}_1^1p \rightarrow {}_0^1n + {}_{+1}^0e + \nu_e$

Quantity	Before interaction $_1^1p$	Total before interaction	After interaction $_0^1n$	$_{+1}^0e$	ν_e	Total after interaction	Conserved?
Q	+1	+1	0	+1	0	+1	✓
B	+1	+1	+1	0	0	+1	✓
L	0	0	0	−1	+1	0	✓
S	0	0	0	0	0	0	✓

All quantities are conserved, so the interaction is permitted.

Exam tip

You will always be given particle quantity values, either as part of the question or on the datasheet.

Summary

■ During all particle interactions, mass-energy, momentum, charge, baryon number and lepton number are always conserved.

■ Strangeness is conserved in strong and electromagnetic interactions, but can change by ±1 or stay the same during weak interactions.

■ Conservation laws can be used to decide whether a particle interaction is allowed or not.

Electromagnetic radiation and quantum phenomena

The photoelectric effect

When photons collide with the surface of materials (usually metals) they can interact with electrons in the atoms at the surface. If the photons have enough energy, the electrons with which they interact can absorb the energy of the photon and escape from the surface – this is called the **photoelectric effect** (see Figure 19). Electrons are held to the surface of the metal with a characteristic amount of energy, called the work function, ϕ, of the metal. If the energy of the incident photon, $E = hf$, is greater than the work function, then the photoelectron emitted will have a maximum kinetic energy, E_{kmax}, given by:

$$E_{kmax} = hf - \phi$$

which can be rearranged to get:

$$hf = \phi + E_{kmax}$$

This equation also leads to the definition of a threshold frequency, f, which is the frequency of the photon that gives the photon enough energy to emit a photoelectron with zero kinetic energy. This frequency is given by:

$$f = \frac{\phi}{h}$$

Photoelectric emission occurs only with photons having a frequency greater than the threshold frequency. If photons of a lower frequency are used, even if their intensity is very high, this will not result in the emission of photoelectrons.

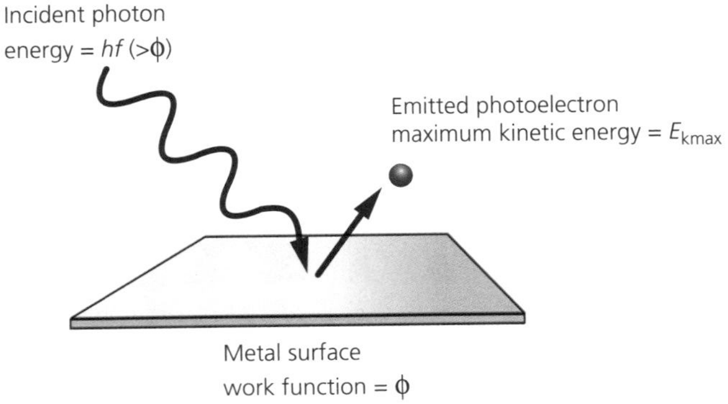

The **photoelectric effect** involves the emission of electrons from a metal surface due to the absorption of photons with energy greater than the work function of the metal.

Knowledge check 25

Photons with energy of 6.5 eV are incident on a metal surface with a work function of 3.0 eV. Calculate the maximum kinetic energy of any emitted photoelectrons.

Incident photon
energy = hf (>ϕ)

Emitted photoelectron
maximum kinetic energy = E_{kmax}

Metal surface
work function = ϕ

Figure 19 The photoelectric effect

Applying a **reverse polarity** potential difference to the surface of the metal will reduce the ability of the emitted photoelectrons to escape from the surface. If the reverse polarity voltage is increased, eventually the maximum kinetic energy of the emitted photoelectrons will not be enough to escape from the surface. This voltage, V_s, is called the stopping voltage.

Reverse polarity of a potential difference means having a potential difference that opposes the flow of electrons.

Worked example

During the photoelectric effect, a gold surface in a vacuum is irradiated with ultraviolet photons of a certain frequency. Photoelectrons are emitted from the gold surface.

a Explain why the emitted photoelectrons have a maximum value.

b Gold has a work function, $\phi = 5.1\,eV$. Explain why, if the frequency of the radiation is below a certain value, photoelectrons are not emitted.

c State another unit for work function.

d The frequency of the light is $1.8 \times 10^{15}\,Hz$. Calculate the energy of an incident photon in eV.

e Calculate maximum energy of the emitted photoelectrons in eV.

Answers

a The law of conservation of energy means that the maximum energy of the photoelectrons is equal to the energy of the photons minus the work function of the surface, $E_{kmax} = hf - \phi$.

b If $hf < \phi$, then there is not enough energy to emit photoelectrons.

c Electronvolts are a unit of energy, so another unit is joules, J.

d $E = hf = 6.6 \times 10^{-34}\,J\,s \times 1.8 \times 10^{15}\,Hz = 1.2 \times 10^{-18}\,J = 7.4\,eV$

e $E_{kmax} = hf - \phi = 7.4\,eV - 5.1\,eV = 2.3\,eV$

Summary

- The photoelectric effect involves the emission of electrons from a metal surface as the result of absorbing photons with enough energy, $E = hf$, to overcome the work function, ϕ, of the metal.
- The photoelectric equation is $hf = \phi + E_{kmax}$, where f is the frequency of the incident photons and E_{kmax} is the maximum energy of the emitted photoelectrons.

Collisions of electrons with atoms, energy levels and photon emission

Electrons within atoms can gain or lose energy as they move within the atom. They have least energy close to the nucleus, called their **ground state**. If an electron gains energy by absorbing a photon it will move further away from the nucleus, and then it can move back down closer to the nucleus by emitting a photon of the same energy. The electrons therefore move up or down in energy level. Electrons can only absorb photons with specific energies because the possible positions, or allowed energies, for electrons in an atom are not continuous. Only certain fixed discrete values are allowed. The electrons can jump from one energy level to another either by absorbing photons or by emitting photons of the same energy as the difference between the

The **ground state** of an electron around an atomic nucleus is its lowest energy level possible.

energy levels. If an electron moves from a high-energy level E_1 to a lower-energy level E_2 it will emit a photon with an energy $E = E_1 - E_2$, so:

$$hf = E_1 - E_2$$

The opposite is also true — if an electron moves from a lower-energy level E_2 up to a higher-energy level E_1 it must absorb a photon equivalent to $hf = E_1 - E_2$.

The allowable energy levels within an atom are usually given energy values in **electronvolts, eV**, where $1\,eV = 1.60 \times 10^{-19}\,J$. All energy levels have negative values — this is because they are all potential energies. The zero of potential energy between two charged particles (such as an electron and a nucleus) is when they are an infinite distance apart, so their energy level values must be negative as energy is absorbed by electrons moving them up to a higher energy level. The energy levels of the electron within the hydrogen atom are shown in Figure 20.

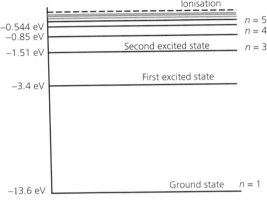

Figure 20 The electron energy levels in the hydrogen atom

The evidence for the discrete energy levels within atoms comes from the observation of line spectra emitted from elements in their gas state at high energy, such as within a star, or inside a gas discharge tube. The light can be split up into its spectrum of different coloured spectral lines by a prism or a diffraction grating. The fact that only certain colours are emitted shows that electron energy levels are discrete and not continuous.

Electrons in energy levels higher than their (empty) ground state are said to be *excited*. If enough energy is absorbed by an electron it can leave the atom altogether. This is called *ionisation* and leaves the atom positively charged. Fluorescent tube lights work by exciting a low-pressure mercury vapour inside the tube using an electric potential difference across the gas. The excited electrons cascade back down to their ground state, emitting ultraviolet photons as they do so. The ultraviolet photons strike a fluorescent coating on the inside of the tube, causing it to glow, giving out the characteristic light of the tube.

The **electronvolt, eV**, is a unit of energy equivalent to $1.60 \times 10^{-19}\,J$. It is frequently used as a convenient unit for describing energies on the microscopic quantum level.

Knowledge check 26

Calculate the wavelength of photons emitted due to an electron jumping from an energy level of $-1.51\,eV$ to an energy level of $-3.4\,eV$.

Exam tip

In examination questions, energy levels may be quoted in J or eV.

Knowledge check 27

Calculate the ionisation energy in joules of a hydrogen electron in its ground state of $-13.6\,eV$.

Knowledge check 28

Explain the difference between excitation and ionisation of electrons.

Worked example

When atoms of hydrogen in their ground state collide with free electrons, the hydrogen atoms can be excited or ionised.

a Explain the difference between excitation and ionisation.

b The ionisation energy of hydrogen is 13.6 eV. Calculate the minimum frequency necessary for a photon to cause the ionisation of a hydrogen atom. Give your answer to an appropriate number of significant figures.

c Figure 21 shows the three lowest energy levels of hydrogen. An electron of energy 10.4 eV is incident on a hydrogen atom. As a result an electron in the ground state of the hydrogen atom is excited. The atom then subsequently emits a photon of a characteristic frequency.

Explain why the electron in the ground state becomes excited to the $n = 2$ energy level.

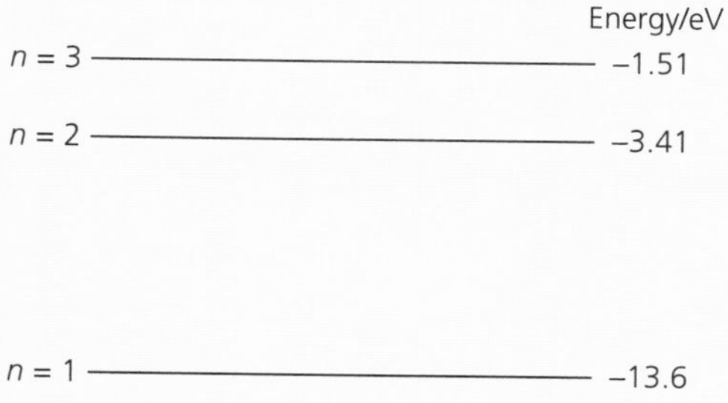

Energy/eV

$n = 3$ ———————————————— -1.51

$n = 2$ ———————————————— -3.41

$n = 1$ ———————————————— -13.6

Figure 21 The three lowest energy levels of hydrogen

d Calculate the wavelength of the emitted photon.

Answers

a During excitation, electrons in the atom are raised to energy levels above their ground state. When an atom is ionised an electron is removed from the atom.

b $E = hf \Rightarrow f = \dfrac{E}{h} = \dfrac{13.6 \text{ eV} \times 1.6 \times 10^{-19} \text{ J}}{6.6 \times 10^{-34} \text{ J s}} = 3.3 \times 10^{15} \text{ Hz}$

c If the electron is initially in its ground state of -13.6 eV, then raising it by 10.4 eV moves it up to -3.2 eV above the $n = 2$, -3.41 eV energy level into which it will drop.

d $hf = E_1 - E_2 = -3.41 \text{ eV} - (-13.6 \text{ eV}) = 10.19 \text{ eV} = \dfrac{hc}{\lambda}$

$\Rightarrow \lambda = \dfrac{6.6 \times 10^{-34} \text{ J s} \times 3 \times 10^8 \text{ m s}^{-1}}{10.19 \text{ eV} \times 1.6 \times 10^{-19} \text{ J}} = 1.2 \times 10^{-7} \text{ m}$

Content Guidance

Summary

- Electrons exist in discrete energy levels and can change energy level by the absorption or emission of photons.
- The energy of emitted/absorbed photons is given by $hf = E_1 - E_2$, where E_1 and E_2 are two electron energy levels.
- Electrons above their lowest (ground state) energy level are said to be excited and if an electron is given enough energy it can leave the atom altogether, which is called ionisation.
- Electron energy levels are usually measured in electronvolts, eV, where $1\,\text{eV} = 1.60 \times 10^{-19}\,\text{J}$.

Wave–particle duality

When a beam of electrons is passed through a wafer-thin target of graphite, the interaction between the electrons and the hexagonal pattern of graphite atoms produces a pattern on a fluorescent screen that closely resembles the diffraction pattern of light that has passed through a (circular) diffraction grating, as illustrated in Figure 22.

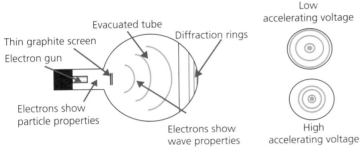

Figure 22 Electron diffraction through a graphite target

This experiment illustrates that in some cases, such as electron diffraction, electrons can behave as waves, and in other cases, such as the photoelectric effect, they can behave as particles. In other words, electrons, like all other particles on the microscopic quantum scale, have a **wave–particle duality**.

The wave behaviour of electrons was first formalised by Louis de Broglie in 1924. de Broglie came up with the relationship between the **momentum** of the photon, mv, and its 'de Broglie' wavelength, λ:

$$\lambda = \frac{h}{mv}$$

Increasing the energy of the electron beam increases the momentum of the electrons and decreases the de Broglie wavelength. The effect of this on the electron diffraction

Knowledge check 29

What does electron diffraction show?

Wave–particle duality describes the contradictory behaviour of particles and photons that behave as waves under certain circumstances and particles under other circumstances.

Momentum is the product of the mass of a particle and its velocity.

Exam tip

The wavelengths of electrons involved with electron diffraction are usually given in nm or pm. Remember to change these to m using the correct standard form before using the numbers in calculations.

pattern is to reduce the distance between the diffraction maximum circles, as shown in Figure 22.

Our knowledge and understanding of matter changed considerably during the first half of the twentieth century and Louis de Broglie was awarded the Nobel Prize for Physics in 1929, but only after his theoretical predictions were confirmed experimentally by Clinton Davisson and Lester Germer in 1927. Changes in the fundamental understanding of physics need to be evaluated through peer review and validated by the scientific community before they can become accepted as fact. The knowledge and understanding of the nature of matter continue to change over time as new theories and experiments continually push the boundaries of our understanding of the universe.

Knowledge check 30

Calculate the de Broglie wavelength of an electron with a momentum of 4.0×10^{-23} kg m s^{-1}.

Worked example

Electrons can behave like waves when they travel through a graphite grid acting as a transmission diffraction grating, provided that the wavelength of the electrons is similar to the atomic spacing of the graphite atoms, 0.14 nm.

a Calculate the speed of electrons with a wavelength of 0.14 nm if the mass of an electron is 9.11×10^{-31} kg.

b Muons have a mass 207 times the mass of an electron and can also undergo diffraction. Calculate the speed of muons with the same wavelength as the electrons in a).

Answers

a Rearranging the de Broglie equation gives:

$$\lambda = \frac{h}{mv} \Rightarrow v = \frac{h}{m\lambda} = \frac{6.6 \times 10^{-34} \text{ J s}}{9.11 \times 10^{-31} \text{ kg} \times 0.14 \times 10^{-9} \text{ m}} = 5.2 \times 10^{6} \text{ m s}^{-1}$$

b $$v = \frac{h}{m\lambda} = \frac{6.6 \times 10^{-34} \text{ J s}}{207 \times 9.11 \times 10^{-31} \text{ kg} \times 0.14 \times 10^{-9} \text{ m}} = 2.5 \times 10^{4} \text{ m s}^{-1}$$

Summary

- On the microscopic, quantum scale particles can behave as waves or particles. This is called wave–particle duality.
- Electron diffraction shows electrons behaving as waves. The photoelectric effect shows waves behaving as particles.
- The de Broglie relationship relates the wavelength, λ, of a particle to its momentum, mv, through $\lambda = h/mv$.
- Increasing the momentum of the particle decreases its wavelength λ. If $\lambda \gg$ dimension of the diffraction source, then the amount of diffraction decreases with increasing momentum.

∎Waves

In **Waves** you need to know about the terms and concepts used to describe progressive and stationary waves. You need to be able to describe the similarities and differences between transverse and longitudinal waves and the principle of superposition of waves. You also need to describe the principles behind interference, diffraction and refraction.

Progressive and stationary waves

Progressive waves

Progressive waves transfer energy from one place to another. In order to move, the waves must have something to move through — this is called a medium (e.g. water waves move through water and sound moves through air). As the waves propagate they cause the particles of the medium to oscillate.

A **progressive wave** carries energy from one place to another.

Describing waves

When describing progressive waves the following terms are used:

- Amplitude is a measure of the energy of the wave. It measures the difference between the maximum displacement of the wave and the undisturbed medium that the wave passes through.
- Wave speed, c, is the rate of motion relative to the medium, measured in metres per second, m s^{-1}.
- Frequency, f, is the number of waves per second propagating through the medium, measured in hertz, Hz. $1\,\text{Hz} = 1$ complete wave per second. Frequency is related to the time period T (measured in seconds) of a wave (the time taken for one wave to repeat itself) through $f = 1/T$.
- Wavelength, λ, is the distance that the wave takes to repeat itself once, measured in metres, m.

Wave speed, frequency and wavelength are related to each other by the wave equation:

$$c = f\lambda$$

Phase and phase difference

As waves repeat themselves with a regular period or cycle, the **phase** of a wave describes the fraction of the wave cycle that has elapsed since the origin of the wave. Phase is usually measured as an angle, either in degrees (where one complete cycle is $360°$) or in radians (where one complete cycle is 2π radians) *or* by the fraction of the cycle. For example, a wave that is a quarter of the time through its cycle has a phase of $90°$, $\frac{\pi}{2}$ radians or $\frac{T}{4}$ fraction.

Two points on a wave, or two different waves, can be compared by their **phase difference**, Φ. This is the difference in the phase angle or fraction of the cycle between the two points. Two points with the same phase are said to be *in-phase* and two points that are exactly half a cycle or $180°$ or π radians apart are said to be in *anti-phase*.

> **Knowledge check 31**
>
> Calculate the frequency of an ultrasound wave with a time period of $40\,\mu s$.

> **Knowledge check 32**
>
> Calculate the frequency of red laser light with $\lambda = 632.8\,\text{nm}$.

The **phase** of a wave describes the fraction of the wave cycle that has elapsed since the origin of the wave.

Phase difference describes the difference in phase angle between two points on a wave or two different waves.

Worked example

Water waves are transverse waves that travel along the surface of a body of water such as a lake. Figure 23 shows the displacement of three buoys on the surface of the lake as one such water wave travels across the surface of the lake.

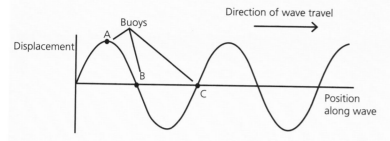

Figure 23 Water waves travelling across the surface of a lake

Buoys B and C are 12.0 m apart and one complete wave passes buoy C in 4.0 s.

a Calculate:

  i   the frequency of the waves

  ii the speed of the waves

b State the phase difference between buoys:

  i   A and B

  ii A and C

c Describe the motion of buoy B during the passage of the next complete wave cycle.

Answers

a **i** $\text{frequency} = \dfrac{1}{\text{time period}} = \dfrac{1}{4.0\text{ s}} = 0.25\text{ Hz}$

  ii wave speed = frequency × wavelength = 0.25 Hz × 12.0 m = 3.0 m s^{-1}

b **i**   A and B are one-quarter of a cycle out of phase, so their phase difference is $\dfrac{\pi}{2}$ radians or 90°.

  ii A and C are three quarters of a cycle out of phase, corresponding to a phase difference of $\left(\dfrac{3\pi}{2}\right)$ radians or 270°.

c The motion of the buoy is at right angles to the direction of energy transfer of the wave. The buoy moves up to a position of maximum displacement, back to equilibrium, then to maximum negative displacement and finally back to the equilibrium position again.

Measuring the speed of sound

The speed of sound in air can be measured directly by using a pair of microphones and a datalogger. The microphones are placed a known distance d apart and a sound is generated close to one of the microphones. The time delay Δt between the two microphones picking up the sound is measured using the datalogger timing program, then the speed of sound is calculated via $c = d/\Delta t$.

Measuring the speed of water waves

Water waves can be produced easily inside a flat trough of water. By lifting the trough slightly and letting it fall, a wave is generated that can propagate from one end of the trough to the other. The speed of the water waves can be determined by timing how long it takes for the waves to travel a measured, known distance, from one end to the other. The speed of the waves can be determined for a range of depths and drop heights.

Summary

- The oscillation of the particles of the medium as a wave passes through can be described in terms of amplitude, frequency, wavelength, speed, phase and phase difference.
- The speed, frequency and wavelength of a wave are related by $c = f \times \lambda$.
- The frequency and time period of a wave are related by $f = 1/T$.
- The phase difference between two waves can be measured as an angle (in radians or degrees) or as fractions of a cycle.

Longitudinal and transverse waves

Longitudinal and transverse waves can be demonstrated on a slinky coil. The wave pulses can travel along the slinky coil as **transverse** waves, where the coils (or particles or electromagnetic field) move at right angles to the direction of travel of the wave in a series of 'crests' and 'troughs', or they can travel as a series of 'compressions' and 'rarefactions' in the same direction as the wave travels along the slinky — **longitudinal** waves. All electromagnetic waves are examples of transverse waves, as are secondary seismic s-waves. Primary seismic p-waves and sound are examples of longitudinal waves. All electromagnetic waves travel at the same speed in a vacuum, $3 \times 10^8 \, \mathrm{m\,s^{-1}}$.

In a **transverse** wave the coils (or particles or electromagnetic field) move at right angles to the direction of travel of the wave. In a **longitudinal** wave the coils or particles move in the same direction as the wave travels.

Polarisation

Transverse waves can be polarised by a suitable polariser. This property can be used to distinguish between transverse and longitudinal waves. Non-polarised (transverse) electromagnetic waves have their electric and magnetic fields propagating in all directions. If the wave is passed through a polariser, the material of the polariser absorbs the components of the fields in the direction of the polariser. Components of the field at right angles to the polariser travel through unaffected, leaving only one plane of polarisation, as shown in Figure 24.

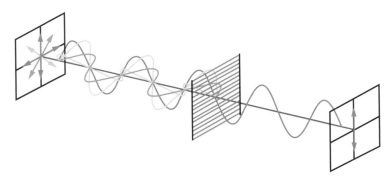

Figure 24 Polarisation of a transverse wave

Polarising sunglasses include lenses with polarising (Polaroid) filters orientated to reduce the glare of light reflecting from horizontal surfaces such as water and snow, thus reducing eye strain for skiers and making vision safer for drivers. Radio and television aerials also work by utilising the polarising effect. Transmitters generate **plane-polarised electromagnetic waves**, which are picked up most effectively by a receiver with the same plane of polarisation.

> **Plane-polarised waves** have particles or a field that always oscillates in the same plane.

> **Knowledge check 35**
>
> Explain why polarisation can be used as a way of determining whether a wave is transverse or longitudinal.

Worked example

Light is an example of a transverse wave. Sound is an example of a longitudinal wave.

a Give one other example of a transverse wave and one other example of a longitudinal wave.

b State one difference and one similarity between a transverse and a longitudinal wave.

c Figure 25 shows unpolarised light from a light bulb passing between two light polarisers, A and B. The arrows show the plane of oscillation of the waves of light. Draw the transmission axis of polariser B.

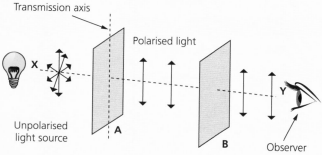

Figure 25 Unpolarised light travelling through two polarisers

d Describe what happens to the intensity of the light perceived by the observer as polariser B is rotated through 180° about line XY.

Answers

a Transverse waves: radio, microwaves, infrared, ultraviolet, X-rays, gamma rays, water waves, rope, transverse slinky waves, seismic s-waves.
Longitudinal waves: ultrasound, seismic p-waves, longitudinal slinky waves.

b Differences: direction of vibration is at right angles to the direction of energy transfer for a transverse wave and is in the same direction for longitudinal waves; transverse waves can be polarised; all longitudinal waves need a medium to travel through, transverse electromagnetic waves do not require a medium.
Similarities: both waves transmit energy; both waves obey $c = f\lambda$.

→

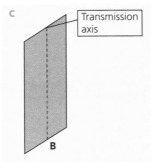

Figure 26 Transmission axis of polariser B

d The intensity falls to a minimum at 90° and then rises again to a maximum at 180°.

Summary

■ Longitudinal waves (such as sound) have their axis of particle displacement co-linear (along the same line) to the direction of energy propagation of the wave. Transverse waves (such as electromagnetic waves and the waves on a string) have their direction of displacement of particles/fields at right angles to the energy propagation of the wave.
■ All electromagnetic waves travel at the same speed in a vacuum, $3.0 \times 10^8\,\mathrm{m\,s^{-1}}$.
■ Transverse waves can be polarised, longitudinal waves cannot.
■ The Polaroid lenses in sunglasses and the alignment of aerials for transmission and reception of television and radio signals are applications of polarisation.

Principle of superposition of waves and formation of stationary waves

The principle of superposition

When two waves of the same type meet at a point they can interfere with each other. Superposition occurs between the two waves, where the displacement of each wave is added as a vector sum producing a resultant wave. If the two waves arrive in-phase, **constructive superposition** occurs and a wave with maximum amplitude is formed. If the waves arrive out-of-phase, destructive superposition occurs and a resultant wave with minimum amplitude is formed. If the two waves have equal amplitudes the resultant wave has double amplitude during constructive superposition and zero amplitude during destructive superposition, as shown in Figure 27.

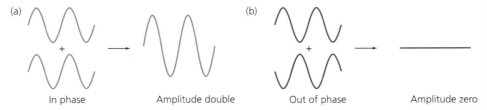

Figure 27 (a) Constructive and (b) destructive superposition

Stationary waves

Stationary waves are formed when two waves of the same frequency travelling in opposite directions superimpose on each other. Look at Figure 28.

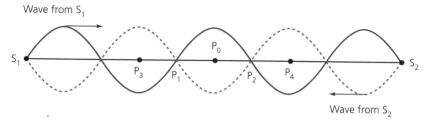

Figure 28 Standing wave formation

Two coherent (in-phase and with the same frequency) waves are generated by sources S_1 and S_2 and travel in opposite directions. At the midpoint, P_0, the two waves always arrive in-phase and constructive superposition occurs and a wave with maximum amplitude is formed, called an **antinode**. At points P_1 and P_2, the waves always arrive out-of-phase, destructive superposition occurs and the waves cancel each other out, forming **nodes**. At points P_3 and P_4, the waves are also in-phase; as the waves are displaced by exactly one wavelength, **constructive interference** occurs and antinodes are formed.

Standing waves can be formed on a vibrating string under tension, as shown in Figure 29.

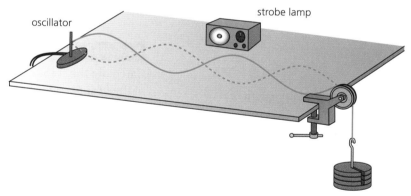

Figure 29 Standing waves forming on a vibrating string

Stationary waves appear not to move and consist of a series of maximum amplitude antinodes and zero amplitude nodes.

Constructive interference occurs when waves arrive in-phase and combine to form a wave with maximum amplitude.

Exam tip

Do not confuse nodes and antinodes. Remember, NOdes have NO amplitude.

Knowledge check 36

The distance between two adjacent nodes on a standing wave of sound inside a trumpet is 15 cm. If the speed of sound in air is $330 \, \text{m s}^{-1}$, what is the frequency of the note played on the trumpet?

For a fixed tension in the string, as the frequency of the vibrating source increases a series of standing waves, called harmonics, is formed, as shown in Figure 30.

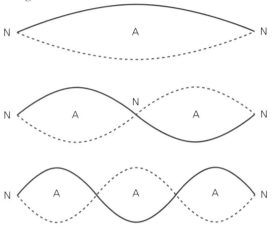

Figure 30 Harmonics on a vibrating string

The lowest-frequency standing wave is called the first harmonic and the frequency is given by the equation:

$$f = \frac{1}{2l}\sqrt{\frac{T}{\mu}}$$

where T is the tension force on the string, l is the length of the string and μ is the mass per unit length of the string.

Worked example

A G electric guitar string of length 640 mm has a mass per unit length of $1.14 \times 10^{-3}\,\text{kg m}^{-1}$ and is put under a tension of 73.2 N.

a Calculate the frequency of the first harmonic standing wave on the string.

b Calculate the speed of a progressive wave on this string.

c The guitarist needs to produce a sound with a higher frequency than the first harmonic on the same string. State one way that this can be achieved.

Answers

a
$$f = \frac{1}{2l}\sqrt{\frac{T}{\mu}} = \frac{1}{\left(2 \times 640 \times 10^{-3}\,\text{m}\right)}\sqrt{\frac{73.3\,\text{N}}{1.14 \times 10^{-3}\,\text{kg m}^{-1}}} = 198\,\text{Hz}$$

b The wavelength of the first harmonic wave is twice the length of the string, $\lambda = 2 \times 640 \times 10^{-3}\,\text{m} = 1.28\,\text{m}$.
Wave speed c = frequency × wavelength = 198 Hz × 1.28 m = 253 m s^{-1}.

c Increase the tension in the string

Exam tip

The wavelength of the first harmonic wave on a string = 2 × length of string.

Knowledge check 37

Calculate the frequency of the first harmonic standing wave on a steel wire of length 1.5 m and mass 3.75 g with a tension of 5.6 N applied to the wire.

Required practical 1

Investigation into the variation of the frequency of stationary waves on a string with length, tension and mass per unit length of the string

The equation for the frequency of the first harmonic standing wave on a string can be used to investigate a vibrating string. The frequency of the first harmonic can be measured from the signal generator driving the vibrating oscillator and the other variables. The length of the string can then be varied by moving the oscillator along the string and the tension can be varied by adding or subtracting masses from the slotted mass holder attached to the string. The mass per unit length can be varied by changing the type of string (this is usually achieved by using metal wires of different gauges).

In one such experiment, a student was trying to measure the mass per unit length μ of a guitar string, 0.64 m long. She measured the first harmonic frequency f using a microphone and an oscilloscope for different tensions T. Her data are shown in the table.

Tension, T/N	First harmonic frequency, f/Hz
0.5	85
1.0	125
2.0	175
3.0	205
4.0	250
5.0	275
6.0	300
7.0	325

The first harmonic frequency can be measured with a resolution of ±5 Hz. The equation linking the first harmonic frequency and the tension is:

$$f = \frac{1}{2l}\sqrt{\frac{T}{\mu}}$$

Squaring both sides of this equation gives:

$$f^2 = \frac{1}{4l^2}\frac{T}{\mu}$$

Plotting a graph of f^2 (y-axis) against T (x-axis) should give a straight line with a gradient equal to $1/4l^2\mu$. Plot this graph and use the spread of the data to determine a value of μ with a ± uncertainty.

For answers see page 92.

Standing waves with microwaves and sound

Standard laboratory apparatus can be used to investigate the standing waves produced by sound and microwaves. The principles of the experiments are similar — waves from a source are reflected back off a suitable reflector and the standing wave pattern can be investigated using a receiving aerial (microwaves) or a microphone (sound) attached to an oscilloscope, as shown in Figure 31. The detector is moved backwards and forwards between the source and the reflector. A ruler or tape measure can be used to mark the positions of nodes and antinodes.

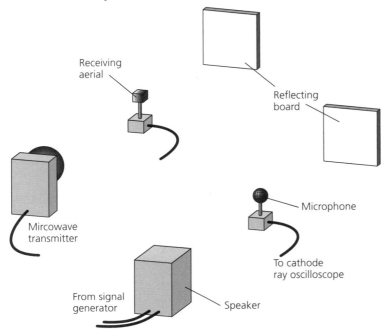

Figure 31 Standing waves formed by microwaves and sound

Exam tip

Microwaves and sound waves are often used as examples of waves in examination questions. Standard laboratory microwaves produced by a school microwave transmitter have a wavelength of about 3 cm — the same as 11 kHz sound waves at sea level.

Summary

- Stationary waves form when two waves of the same frequency travel in opposite directions and interfere with each other.
- Stationary waves are described in terms of nodes (zero amplitude) and antinodes (maximum amplitude).
- Stationary waves on strings can be described in terms of harmonics. The first harmonic of a wave on a string is defined by:

$$f = \frac{1}{2l}\sqrt{\frac{T}{\mu}}$$

- Stationary waves can be easily formed in the laboratory: the vibration of strings, the reflection of microwaves and the reflection of sound.

Refraction, diffraction and interference

Interference

Coherence

Two wave sources are said to be **coherent** if they:

- produce waves of the same type (e.g. water waves, light of the same wavelength etc.)
- produce waves with the same frequency
- produce waves with the same phase, or there is a constant phase difference between the wave produced by each source.

Two waves are **coherent** if they are of the same type, have the same frequency and are in-phase or have a constant phase difference.

Two dippers attached to the same ripple tank source are coherent sources, as are the monochromatic (single-frequency) photons of light produced by a laser. The polychromatic (multi-frequency) white light produced by a light bulb cannot be coherent as a range of different frequencies and phase differences are produced. The coherent photons of light emitted by a laser are usually produced within a narrow beam within a small area. This produces light with a very high intensity, even though the power of the laser is usually only a couple of milliwatts (for a standard school laser).

The intensity of light produced by a laser could damage the retina inside your eye, so special precautions are usually required when using lasers:

- The intensity can be reduced by using a neutral density filter that absorbs some of the photons.
- Laser goggles can be worn.
- Eyes should be kept away from the plane of the beam and unwanted reflections should be avoided.
- A 'laser in operation' sign should be put up on the door of the laboratory.

Path difference

Two, in-phase, coherent sources S_1 and S_2 produce waves that propagate away from the sources towards points O, P and Q, as shown in Figure 32.

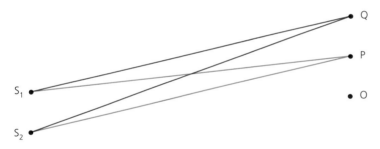

Figure 32 The path difference between two waves

The distances S_1O and S_2O are the same, in other words, the **path difference** $|S_2O - S_1O| = 0$. This means that waves from S_1 and S_2 arrive at O in-phase and constructive superposition occurs. At position Q, if the path difference $|S_2Q - S_1Q| = \lambda$, one whole wavelength (i.e. the waves travelling from S_2 have travelled a distance λ, more than the waves travelling from S_1), then the waves will also arrive in-phase and constructive superposition will occur once again. However, at position P, if the path difference is half a wavelength, or $|S_2P - S_1P| = \frac{\lambda}{2}$, then the waves arrive out-of-phase and destructive superposition occurs. This pattern can be repeated — wherever the path difference is zero or a whole number of wavelengths, constructive superposition occurs, and wherever the path difference is an odd number of half wavelengths, destructive superposition occurs.

If the **path difference** between two coherent waves is equal to a whole number of wavelengths then constructive superposition occurs. If the path difference is an odd number of half wavelengths then destructive superposition occurs.

Young's double slit interference and the nature of light

In 1804, Thomas Young published a paper describing his experiments on the diffraction and interference of light. Young was convinced that light behaved as a wave (despite Newton's 'corpuscular' (particle) model of light published 100 years earlier in 1704). So he carried out a series of experiments involving the diffraction of light through a double slit, in an effort to prove the wave theory originally suggested by Christiaan Huygens in 1678. Figure 33 shows a modern version of Young's double slit experiment.

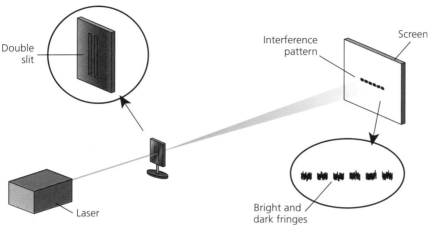

Figure 33 Young's double slit experiment

The interference pattern set-up on the screen shows a series of bright and dark fringes – the bright fringes occur where the path difference between the light arriving from the two slits is a whole number of wavelengths and constructive superposition occurs, and the dark fringes occur where the path difference is an odd number of half wavelengths and **destructive interference** occurs. If the fringe spacing is given as w, the wavelength of the light λ, the distance from the slits to the screen D, and the slit separation s, then:

$$w = \frac{\lambda D}{s}$$

As interference is a property of waves, Young confidently underlined the wave nature of light, and this model was to prevail until the turn of the twentieth century with the work done by Max Planck, Albert Einstein, Louis de Broglie and others, who formalised the wave–particle duality of light, laying the foundations of modern quantum physics in the process.

Worked example

In 1804, Thomas Young published his results from an experiment illustrating the interference of polychromatic light from the Sun passing through two close slits cut into a sheet of cardboard.

a   Explain why Young's conclusions about this experiment were criticised by supporters of Newton's 'corpuscular' theory of light.

b   This experiment is best demonstrated using the **monochromatic, coherent light** from a standard laboratory laser.

   i   State one safety precaution that should be taken when using a laboratory laser.

   ii   State what is meant by monochromatic light.

   iii State what is meant by coherent light.

➜

Destructive interference occurs when waves arrive out-of-phase and cancel each other out, forming a wave of minimum amplitude.

Monochromatic light has a single wavelength (colour), whereas polychromatic light is made up of a spectrum of different wavelengths (colours).

c In one such experiment involving a red laser, the red fringes are measured to be 1.3 cm apart, on a screen 4.6 m from the double slits. The slit separation is 0.22 mm. Calculate the wavelength of the light emitted from the laser.

d The experiment in c) is repeated using light from the Sun, replicating the experiment reported by Young in 1804. Describe the differences in the patterns observed on the screen.

Answers

a Newton's 'corpuscular' theory of light considered light to consist of a stream of tiny light particles. Young's experiment showed that light behaves as a wave during interference through a double slit.

b i Suitable safety precautions include the following:
  - The intensity can be reduced by using a neutral density filter that absorbs some of the photons.
  - Laser goggles can be worn.
  - Eyes should be kept away from the plane of the beam and unwanted reflections should be avoided.
  - A 'laser in operation' sign should be put up on the door of the laboratory.

ii Only one wavelength.

iii Two light waves are coherent if they are of the same type, have the same frequency and are in-phase or have a constant phase difference.

c $\lambda = \dfrac{w \times s}{D} = \dfrac{0.22 \times 10^{-3} \text{ m} \times 1.3 \times 10^{-2} \text{ m}}{4.6 \text{ m}} = 620 \text{ nm}$

d The monochromatic red laser light pattern will consist of a series of red fringes with a constant spacing of 1.3 cm. The polychromatic white light will produce a series of coloured spectra, the red light of the spectra will have a similar fringe spacing of 1.3 cm, whereas the bluer light will have a lower fringe spacing as w is proportional to the wavelength of the light.

The use of **polychromatic white light** as the source of Young's slits experiment produces a series of 'rainbow' maxima, as shown in Figure 34.

Figure 34 Young's double slit diffraction: comparison between monochromatic (green) light and polychromatic, white light

Knowledge check 39

A green light laser is shone through a pair of slits with a separation of 0.25 mm. A diffraction pattern is observed on a screen 3.2 m from the slits, with a fringe spacing of 6.8 mm. Calculate the wavelength of the laser.

Polychromatic light is not coherent and the different colours produce constructive superposition in different places due to differences in the wavelengths of the different colours.

Knowledge check 40

Explain why a polychromatic light source produces a series of spectral diffraction fringes when shone through a pair of double slits.

Content Guidance

Young's double slit experiment can be replicated with any form of waves. In the laboratory it is convenient to demonstrate it using sound and microwaves, as shown by Figures 35 and 36.

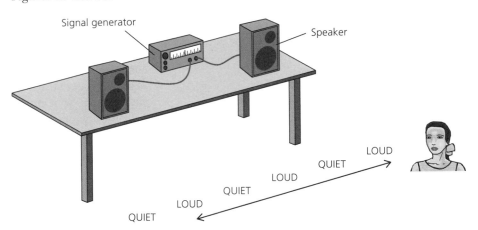

Figure 35 Young's double slit experiment using sound

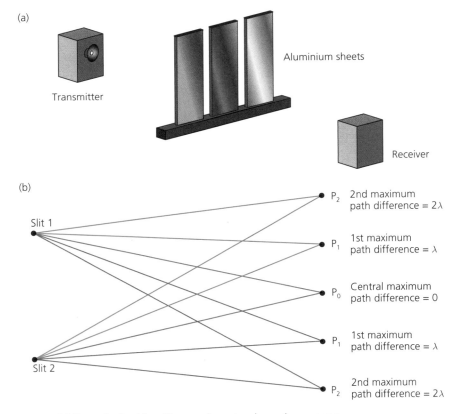

Figure 36 Young's double slit experiment using microwaves

Figure 36 b) shows how the series of microwave maxima is set up using the concept of path difference and constructive superposition where the path differences are zero or a whole number of wavelengths.

Summary

- The path difference between the waves arriving from two sources is the extra distance that one wave has to travel compared with the other.
- Two wave sources are coherent if the waves emitted by them are of the same type, have the same frequency and are emitted in-phase (or with a constant phase difference).
- The monochromatic light from a laser can be used to show interference and diffraction effects in the laboratory.
- In Young's double slit experiment two coherent sources or a single source with double slits can produce an interference pattern, which can be described in terms of $w = \lambda D/s$ where the fringe spacing is w, the wavelength of the light is λ, the distance from the slits to the screen is D, and the slit separation is s.
- White light produces an interference pattern that consists of a spectrum of colours.
- Standard laboratory lasers can cause retinal damage and need to be used safely.
- Our knowledge and understanding of the nature of electromagnetic radiation have changed over time, with light (in particular) being described in terms of a wave model or a particle model and now in terms of wave–particle duality.

Diffraction

Single slit diffraction

Diffraction is a property of waves, causing them to bend around corners or through apertures, as shown by water waves in a ripple tank — see Figure 37.

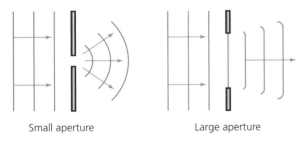

Small aperture   Large aperture

Figure 37 Diffraction of water waves through an aperture

Diffraction occurs when waves travel through a gap or around a barrier.

Exam tip

You do not need to be able to interpret graphs of intensity against angular separation for diffraction patterns.

The amount of diffraction depends on the width of the aperture, being largest when the width of the aperture s is approximately the same size as the wavelength of the waves. Using a laser and an adjustable slit, the width of the central bright maximum increases as the width of the slit gets closer to the wavelength of the laser light. Light of different wavelengths (and colours) produces similar patterns, showing more or less diffraction, depending on the wavelength. For the same slit width, where $s > \lambda$ for both red and blue light, the central diffraction maximum will be wider for red light and narrower for blue light. A polychromatic white light will produce a central bright, white maximum and a series of fainter rainbow-spectra secondary (and higher) maxima.

Diffraction gratings

A (transmission) diffraction grating is a series of extremely close, equally spaced, multiple slits, usually described by the number of slits (or lines) per mm. A good quality laboratory diffraction grating may have 600 lines per mm, which produces a slit separation d of 1×10^{-3} mm/600 $= 1.67 \times 10^{-6}$ m $= 1.67$ μm.

If a diffraction maximum occurs at an angle θ to the direction of the laser beam, then the relationship between the angle, the slit separation d, the wavelength λ, and the maximum order number n, is given by:

$$d \sin \theta = n\lambda$$

Diffraction gratings are used in a variety of applications. Spectrometers, used to analyse the spectra of light coming from stars and galaxies, use diffraction gratings, as do CDs and DVDs, which act as diffraction gratings as a by-product of their manufacture.

Worked example

Blue light from a laser produces a diffraction pattern on a screen 7.6 m from a transmission diffraction grating with 625 lines per mm.

a Calculate the line spacing of the diffraction grating.

b The first order diffraction maximum is observed at an angle of 12° to the normal of the diffraction grating. Use these data to determine the wavelength of the laser.

c Calculate the distance between the centres of the two first order maxima observed on the screen.

d The blue light is replaced by a red laser. Without calculation, state and explain what happens to the diffraction pattern observed on the screen.

Answers

a The line spacing of the grating is calculated from:

$$d = \frac{1 \times 10^{-3} \text{ m}}{625} = 1.6 \times 10^{-6} \text{m}$$

b Rearranging the diffraction grating formula:

$$d \sin \theta = n\lambda \Rightarrow \lambda = \frac{d \sin \theta}{n} = \frac{1.6 \times 10^{-6} \text{ m} \times \sin 12°}{1} = 350 \text{ nm}$$

c The distance between the centres of the two first order maxima $= 2 \times 7.6 \text{ m} \times \tan 12° = 3.2 \text{ m}$.

d Red light has a longer wavelength than blue light. Increasing the wavelength increases $\sin \theta$. If $\sin \theta$ increases, then θ increases, causing the diffraction pattern to spread out.

Knowledge check 41

A laser of wavelength 623 nm produces a first order diffraction maximum at an angle of 11° to the normal line of a diffraction grating. Calculate the number of lines per mm of the grating.

Exam tip

You do not need to describe how a spectrometer works.

Required practical 2

Investigation of interference effects to include the Young's slit experiment and interference by a diffraction grating

A student set up an experiment to compare the diffraction patterns produced by a helium–neon laser, a double slit and a diffraction grating.

The wavelength of the laser is 632.8 nm and the screen is 2.3 m away from the slits/grating.

The pattern produced by the laser beam and the double slit is shown in Figure 38, together with a centimetre ruler scale.

Figure 38 Diffraction pattern of a double slit

The pattern produced by the laser beam and the diffraction grating is shown in Figure 39 with the same centimetre ruler scale.

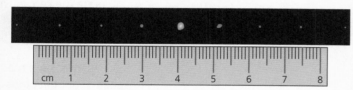

Figure 39 Diffraction pattern of a diffraction grating

The equation for Young's double slit diffraction is given by:

$$w = \frac{\lambda D}{s}$$

where w is the fringe spacing of the diffraction pattern on the screen, λ is the wavelength of the laser light, D is the distance between the screen and the double slits and s is the slit spacing.

The equation for a diffraction grating is given by:

$$d \sin \theta = n\lambda$$

where d is the grating spacing, n is the order number of the diffraction spots and θ is the angle between the 0 order spot, the nth spot and the grating, and λ is the wavelength of the laser light.

Use these equations, the data and the diffraction patterns to compare the slit spacing of the double slits and the grating spacing of the diffraction grating.

For answers see page 93.

Summary

- The diffraction pattern from a single slit using monochromatic light consists of a bright central maximum with much lower intensity secondary maxima. White light produces a similar pattern but exhibits spectra of different colours.
- For a single slit, if the slit width ≈ wavelength of the light, then the width of the central maximum decreases with increasing width and increases with increasing wavelength.
- The diffraction of light through a transmission diffraction grating can be described by the equation $d\sin\theta = n\lambda$.
- Diffraction gratings are used in spectrometers to analyse the light coming from distant stars or galaxies.

Refraction at a plane surface

Refraction occurs when waves travel from one medium, where they travel at speed c_1, into another medium where they travel at a different speed, c_2. The change of speed causes a change in wavelengths and if the waves are travelling at an angle to the boundary between the two media, the wave will appear to change direction, as shown in Figure 40, where water waves in a ripple tank travel from deep water into shallow water.

Refraction is a general property of waves, where the speed of a wave changes when it passes from one medium into another.

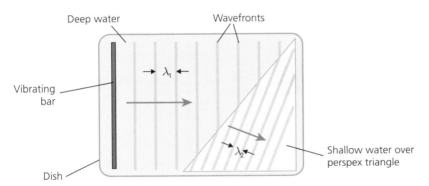

Figure 40 Refraction of water waves in a ripple tank

For the refraction of light through optically transparent substances, the property of the substance that dictates the change of speed is called the refractive index n of the substance, which is given by the equation:

$$n = \frac{c}{c_s}$$

where c is the speed of light in a vacuum and c_s is the speed of the light in the substance. For air, where the speed of light is close to the speed of light in a vacuum, the refractive index is approximately 1.

The relationship between the angle of incidence and the angle of refraction for a refracting wave is given by Snell's law of refraction:

$$n_1 \sin\theta_1 = n_2 \sin\theta_2$$

where n_1 and θ_1 are the refractive index and the angle of incidence in medium 1 and n_2 and θ_2 are the refractive index and the angle of refraction in medium 2, as shown in Figure 41.

Knowledge check 42

Calculate the refractive index of a glass block if the speed of light in a vacuum is $3.0 \times 10^8\,\mathrm{m\,s^{-1}}$ and the speed of light in glass is $2.0 \times 10^8\,\mathrm{m\,s^{-1}}$.

The **refractive index** of a substance is the ratio of the speed of light in the substance to the speed of light in a vacuum.

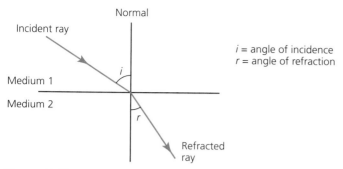

Figure 41 The refraction of light through a boundary

A beam of light travelling through water ($n = 1.3$) hits a glass block ($n = 1.5$) at an angle of 35° to the normal of the boundary. Calculate the angle of refraction inside the glass block.

Total internal reflection

When light travels from a medium with a high refractive index, such as glass, into a medium with a low refractive index, such as air, the refracting beam will refract away from the normal line at 90° to the boundary at the point of refraction, as shown in Figure 42(a).

A beam of light travels through a glass viewing window ($n = 1.5$) looking into the water of an aquarium ($n = 1.3$). The beam hits the glass–water boundary at an angle of 59°. Does the beam travel into the water?

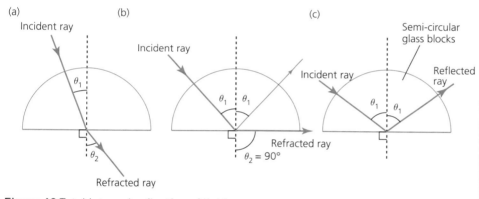

Figure 42 Total internal reflection of light

As the angle of incidence to the boundary θ_1 increases, eventually an angle of incidence is achieved where the refracting beam refracts at an angle of 90° to the normal, along the boundary between the two media, as shown in Figure 42(b). The angle of incidence at this point is called the **critical angle** of the medium, θ_c. For angles of incidence greater than the critical angle, total internal reflection is said to occur and the beam does not refract through the boundary but is reflected back into the medium, as shown in Figure 42(c). The critical angle between two media is given by:

$$\sin\theta_c = \frac{n_2}{n_1}$$

where n_1 and n_2 are the refractive indices of medium 1 and medium 2.

The **critical angle** θ_c is the angle of incidence for which total internal reflection just takes place, and the angle of refraction is 90°.

Fibre optics

One of the most important applications of total internal reflection is its use in fibre optics, where signals of light or infrared are passed down thin strands of transparent materials such as glass and plastics. The beams totally internally reflect off the inside surfaces of the fibres, which are normally coated with a thin outer 'cladding'. This has a lower refractive index than fibre, which ensures total internal reflection, provided that the incident beam is introduced into the fibre at an angle greater than its critical angle, as shown in Figure 43.

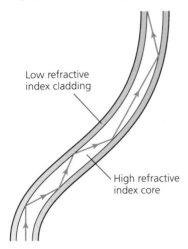

Low refractive index cladding

High refractive index core

Figure 43 Fibre optics

Knowledge check 45

Explain why optic fibres are coated with a (plastic) cladding.

Material and modal dispersion

A spectrum seen using a prism is caused by dispersion. Different colours of light travelling through the glass slow down by different amounts because the refractive index varies with wavelength and the different wavelengths (or colours) travel along slightly different paths. A similar effect occurs inside optical fibres. As a signal travels within an optical fibre, it disperses. Two types of dispersion occur in step-index optical fibres:

■ **Material dispersion** occurs because the refractive index of the optical fibre varies with wavelength. If the signal consists of a sharp pulse of light containing slightly different wavelengths, then the different wavelengths travel at slightly different speeds, which causes the sharp pulse to spread into a broader signal. Therefore the duration of each pulse increases — this is called **pulse broadening**. Pulse broadening is a problem because it limits the maximum frequency of pulses and therefore the bandwidth available for use in the fibre. Ensuring that the pulse contains only monochromatic light (or infrared) reduces this effect substantially.

■ **Modal dispersion** occurs when rays inside the optical fibre take slightly different paths. Rays taking longer paths take longer to travel through the fibre, so the duration of the pulse increases and the pulse broadens again. Narrow fibres reduce this effect.

Absorption

Some wavelengths of light (or infrared) are absorbed strongly by the materials used to make optical fibres, so the signal strength falls. Optical fibres are therefore manufactured from materials with low absorption at the wavelength used to send the signal.

Worked example

An optical fibre is used to send rays of light down through an endoscope, which is used to perform keyhole surgery. A cross-section through part of the fibre is shown in Figure 44.

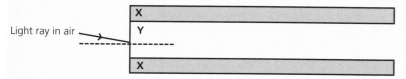

Figure 44 Cross-section through an optical fibre

a   State the name of the parts labelled X and Y on the optical fibre.

b   On a copy of Figure 44 draw the path of the ray of light as it passes through the air–fibre boundary and then undergoes total internal reflection at the boundary between the parts labelled X and Y.

c   The refractive index of part X is 1.51 and the refractive index of part Y is 1.55. Use these data to calculate the critical angle for the boundary between X and Y.

d   Optical fibres are generally made so that the part labelled Y is as narrow as possible. State and explain one reason why this is the case.

e   State one other application of optical fibres (apart from the endoscopes used in keyhole surgery) and explain how this application has benefited society.

Answers

a   X is the cladding, Y is the core of the optical fibre.

b   The ray bends towards the normal on entry into the fibre and then undergoes total internal reflection when it hits the core–cladding boundary.

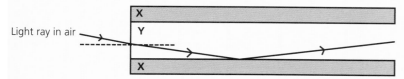

c   The critical angle θ_c is given by:

$$\sin\theta_c = \frac{n_2}{n_1} \Rightarrow \theta_c = \sin^{-1}\left(\frac{n_2}{n_1}\right) = \sin^{-1}\left(\frac{1.51}{1.55}\right) = 77°$$

d   Narrow fibres reduce modal dispersion, which occurs when rays inside the optical fibre take slightly different paths. Rays taking longer paths take longer to travel through the fibre, so the duration of the pulse increases and the pulse broadens again.

e   Use: telephone/cable television/internet communications. Fast, high-volume data communications allow faster/higher-resolution/real-time communications.

Content Guidance

Summary

- The refractive index of a substance is given by $n = c/c_s$.
- The refractive index of air is approximately 1.
- Snell's law of refraction for a boundary is given by $n_1 \sin\theta_1 = n_2 \sin\theta_2$.
- Total internal reflection of a wave occurs when the angle of incidence of the wave is greater than the critical angle θ_c for the interface between two media, where $\sin\theta_c = n_2/n_1$.
- Fibre optics work by using the total internal reflection of light down a fibre made of glass or plastic coated in a material with a higher refractive index called a cladding.
- The signals sent down optic fibres can suffer from absorption, material dispersion — where the refractive index of the fibre varies with the wavelength of the signal — and modal dispersion — where the rays inside the optic fibre take slightly different paths. The effect of these is to broaden the signal pulses and reduce the bandwidth of signals that can be used on the fibre.

Questions & Answers

Sections 1, 2 and 3 are assessed for AS and A-level. There are two AS papers and the content of these sections is assessed on both papers 1 and 2. Both exams are 1 hour 30 minutes long, worth 70 marks in total and are 50% of the AS grade. Paper 1 consists of short and long answer questions split by topic. Paper 2 consists of three sections: section A, worth 20 marks, has short and long answer questions on practical skills and data analysis; section B, worth 20 marks, has short and long answer questions from across all areas of the AS content (including this guide); section C, worth 30 marks, has 30 multiple-choice questions.

There are three A-level papers. Paper 1 includes the content covered in this guide. Paper 2 assesses the content from other sections, but assumes knowledge assessed on paper 1. Both papers are 2 hours long, worth 85 marks, contain short, long and multiple-choice questions, and are 34% each of the A-level grade. Paper 3 consists of two sections: section A assesses practical skills and data analysis and section B assesses the optional sections. Paper 3 is 2 hours long, worth 80 marks and is 32% of the A-level grade. Optional sections assume knowledge from the rest of the specification.

The following two tests are made up of questions that are similar in style and content to those in both the AS and A-level examinations. There are 20 questions in both tests. The first ten questions are multiple-choice, the remaining ten are a mixture of short and longer answer questions. The number of marks for each question is indicated next to the question. You should spend approximately 10 minutes on the multiple-choice questions, short answer questions should take you approximately 5–6 minutes, and longer questions should take approximately 10–15 minutes.

Although these sample questions resemble actual examination scripts, be aware that during the examination you will be writing directly onto the examination paper, which is not possible for this book. It may also be the case that you will need to copy diagrams and graphs that normally you would just write or draw on in the real examination. If you are doing these tests under timed circumstances, you need to allow extra time for this.

Each question has some form of exam guidance. This could be hints on how to answer the questions, a clear statement about what the question is about, suggested content to revisit, or it could take the form of help on the mark schemes, pointing out common mistakes or giving suggestions on things to learn. The student answers give the correct answers to the questions and indicate where the marks are awarded (✔). Many of the answers give guidance showing the sorts of answers given by C/D-grade students (indicated as C-grade) and A/B-grade students (indicated as A-grade).

All the questions in these tests are suitable for AS and A-level examinations.

■ Test paper 1

Multiple-choice questions

For questions 1 to 10 select one answer from A to D.

Question 1

Which of the following particles is a nucleon?

A Electron

B Neutron

C Muon

D Pion

Question 2

Which of the following statements is true about hydrogen, deuterium and tritium, the three isotopes of hydrogen?

A They all have one neutron.

B They have the same nucleon number.

C They have different numbers of electrons.

D They have the same proton number.

Question 3

Which of the following is a unit for the momentum of an electron?

A $J K^{-1}$

B $J kg^{-1}$

C $N kg^{-1}$

D $N s$

Question 4

What is the quark structure of an antiproton?

A $\overline{u}\overline{d}\overline{d}$

B uud

C $\overline{u}\overline{u}\overline{d}$

D udd

Question 5

Figure 1 shows two spinning disks with arrows drawn on them:

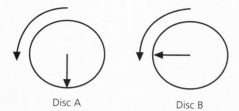

Disc A   Disc B

Figure 1

What is the phase difference between the arrows shown on disc A and disc B?

A 2π radians

B $\pi/2$ radians

C π radians

D $3\pi/2$ radians

Question 6

Below is a list of four graphs that could be plotted relating the speed of a wave c to the wavelength of the wave λ. Which graph would you plot to obtain a straight line through the origin?

A c against λ

B c against λ^2

C c^2 against λ

D c^2 against λ^2

Question 7

Below is a list of four numbers. Which number is the best estimate for the wavelength of visible light, expressed in nm?

A 0.5

B 5

C 50

D 500

Question 8

Figure 2 shows part of the spectrum of mercury. Which letter corresponds to the wavelength with the lowest photon energy?

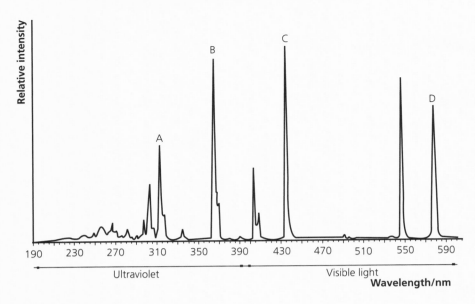

Figure 2

Question 9

Which of the following exchange particles is involved with positron emission?

A W^-

B Photon

C W^+

D Gluon

Question 10

Seismic p-waves travel from limestone where they have a speed of $6100\,\text{m s}^{-1}$ into granite where they travel at $5200\,\text{m s}^{-1}$. What is the refractive index of this boundary?

A 0.85

B 1.17

C 1.20

D 0.83

Answers to multiple-choice questions

1 B Muons and electrons are leptons and a pion is a meson.

2 D Isotopes have the same number of protons (and electrons if they are in atomic form) but different numbers of neutrons.

3 D Ns expressed in fundamental units is $\text{kg m s}^{-2}\text{s}$, which is equal to kg m s^{-1} — this is a mass × velocity = momentum.

4 **C** Antibaryons are made up of three antiquarks.

5 **B** The two arrows are a quarter of a cycle apart, so $2\pi/4 = \pi/2$ radians.

6 **A** $c = f\lambda$ so c is proportional to λ, hence this graph gives a straight line through the origin.

7 **D** The wavelength of visible light runs from about 350 nm up to about 700 nm.

8 **A** $E = hf$ or $E = hc/\lambda$ so the lowest wavelength will have the highest photon energy.

9 **C** Positron emission is a weak interaction. Remember, beta *minus* (electron) emission involves the W *minus*, and beta plus (positron) emission involves the W *plus* exchange particle.

10 **B** Refractive index $n = \dfrac{c_{\text{incident medium}}}{c_{\text{refracted medium}}} = \dfrac{6100\,\text{m s}^{-1}}{5200\,\text{m s}^{-1}}$

Short and longer answer questions

Question 11

During beta minus decay a neutron decays into a proton. The Feynman diagram for beta minus decay is shown in Figure 3.

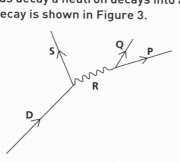

Figure 3

(a) State the names of the particles labelled P, Q, R and S in the diagram above. (4 marks)

(b) Which particle, P, Q, R or S, is an exchange particle? (1 mark)

(c) Name the interaction involved with this decay. (1 mark)

(d) Explain why one of the emitted particles must be an antilepton. (2 marks)

(e) Explain why the beta particles emitted by these decays are emitted with a range of different kinetic energies. (2 marks)

ⓔ You need to remember that during weak interactions, if the decaying particle is positively charged, the exchange particle will be a W^+, and if the decaying particle is negatively charged, the exchange particle will be a W^-.

Student answer

(a) P/Q = electron, e^-/anti-electron neutrino, $\overline{\nu}_e$ (either way around) ✓✓;
R = W^- exchange particle ✓; S = u quark ✓.

🅔 A C-grade student may remember that the other particle produced during the decay of the W⁻ exchange particle is a neutrino, but may not realise that it has to be an antineutrino in order to conserve lepton number.

(b) R (W⁻ exchange particle) ✓.

(c) Weak decay/interaction ✓.

(d) Beta minus decay produces an electron, e⁻, which has a lepton number of $L = +1$ ✓. The down quark has $L = 0$, so one of the particles must be an antilepton, $L = -1$, to ensure conservation of lepton number ✓.

(e) As an electron and an antineutrino are emitted from the decaying neutron, the total energy released as kinetic energy is shared between the two particles ✓, allowing for a range of different values for both ✓.

🅔 A-grade students will get both marks for (d) and (e) because they will state and qualify their answers.

Question 12

A potassium photoelectric cell emits photoelectrons with a maximum kinetic energy of 0.30 eV when monochromatic light is shone onto the potassium surface. The work function of potassium is 2.3 eV.

(a) State the equation that links the photon energy, the work function of the surface and maximum kinetic energy of the emitted photoelectrons. (1 mark)

(b) State what is meant by potassium having a work function of 2.3 eV. (2 marks)

🅔 Although this is a *state* question, there are 2 marks available, so you will need to give two different points — 1 mark will be related to the term *work function* and the other will be related to the *value* 2.3 eV.

(c) Use your answer to (a) and the data in the question to calculate the energy of the monochromatic light photons. (1 marks)

(d) Hence calculate the wavelength of monochromatic light photons. (3 marks)

🅔 Be careful with questions involving non-SI units. In this case the energies are given in eV, but the wavelength will need to be stated in m. You will need to convert energies in eV to joules. You *must* learn how to do this.

(e) Explain why some of the emitted photoelectrons will have a lower kinetic energy than 0.30 eV. (2 marks)

(f) State and explain the effect on the maximum kinetic energy of the emitted photoelectrons of increasing the intensity of the monochromatic light source. (2 marks)

Student answer

(a) $hf = \phi + E_{kmax}$ ✓.

🅔 This equation is on the datasheet so you do not have to remember it. However, you do need to know where to find it. If you are stuck for an equation always check out the datasheet first.

(b) Photoelectrons must have a kinetic energy greater than 2.3 eV ✓ in order to leave the surface of the potassium ✓.

(c) $hf = 2.3\,\text{eV} + 0.30\,\text{eV} = 2.6\,\text{eV}$ ✓.

(d) Convert 2.6 eV to J $= 2.6\,\text{eV} \times 1.6 \times 10^{-19}\,\text{J} = 4.16 \times 10^{-19}\,\text{J}$ ✓. So energy of photon

$$E = hf = \frac{hc}{\lambda} \Rightarrow \lambda = \frac{hc}{E}\ ✓✓ = \frac{6.6 \times 10^{-34}\,\text{J s} \times 3 \times 10^{8}\,\text{m s}^{-1}}{4.16 \times 10^{-19}\,\text{J}}$$

$$= 4.76 \times 10^{-7}\,\text{m},\ 4.8 \times 10^{-7}\,\text{m}\ (2\ \text{sf}) ✓$$

ⓔ C-grade students will score well on the straightforward parts (a) to (d). Error carried forward will operate here.

(e) The work function is minimum energy that photoelectrons are held to the surface ✓ — some will be held with values in the range of 2.3 eV to the energy of the photon and hence can also be emitted ✓. Electrons bound with an energy greater than that of the photon cannot be emitted.

(f) Increasing the intensity of the monochromatic light has no effect on the maximum kinetic energy of the emitted photoelectrons ✓. This is because the energy of the photons is dependent on their frequency, not their intensity ✓.

ⓔ A-grade students will be able to explain both these points in detail.

Question 13

Figure 4 shows some of the electron energy levels in hydrogen.

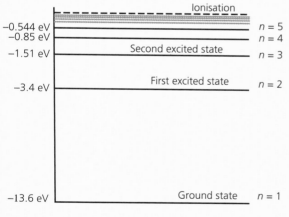

Figure 4

This question is about the absorption and emission of energy when electrons move between energy levels.

(a) An electron, normally in the ground state, can be *excited* or *ionised*. Explain the difference between these two terms.

(2 marks)

(b) Photons with energy of 2.0 × 10^{-18} J cause the electron in its ground state to become excited.

    (i) Calculate the energy of the photons in electronvolts.   (1 mark)

    (ii) State and explain which excited state the electron will move up to after absorbing one of these photons.   (2 marks)

e Remember that electrons can only have energy values equal to the energy levels. If you work out an energy that is greater than a lower energy level value but lower than the next highest energy level, the electron will lose energy and drop back into the lower energy level.

    (iii) Calculate the energy, in electronvolts, of the lowest energy photon that can be emitted when the electron moves back down to its ground state.   (2 marks)

Student answer

(a) Excited means that the electron is moved to an energy level above the ground state, but is still confined to the atom ✓. Ionised means that the electron has an energy greater than 13.6 eV (in this case), removing it from the energy levels of the atom altogether ✓.

e An A-grade student will add the qualifiers to each of the statements above, rather than simple statements of meaning of each of the terms.

(b) (i) $E_{eV} = \dfrac{E_J}{1.6 \times 10^{-19} \, \text{J eV}^{-1}} = 12.5 \, \text{eV}$ ✓

    (ii) $-13.6 \, \text{eV} + 12.5 \, \text{eV} = -1.1 \text{eV}$ ✓ .So electron can reach $n = 3$ ✓, but not $n = 4$.

    (iii) Lowest energy transition available = $n = 3$ to
        $n = 2$ ✓ $= -1.51 \, \text{eV} - (-3.4) \, \text{eV} = 1.89 \, \text{eV}$ ✓

Question 14

A transmission diffraction grating is used to determine the wavelength of a laser. A monochromatic, coherent single ray of light hits the diffraction grating normally and produces a diffraction pattern on a screen.

(a) Describe three features of the diffraction pattern observed on the screen.   (3 marks)

e There will be more than three features that you could describe. Pick the most obvious ones.

(b) Use the principle of superposition to explain the change in the intensity of the diffraction pattern seen on the screen.   (2 marks)

e When you are describing superposition remember that there are two types, constructive and destructive. 1 mark will be for each.

(c) The figure below shows the geometry of the diffraction grating experiment.

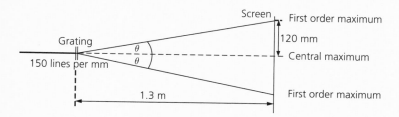

Use data from the figure to show that the grating spacing is about 7 μm.

(1 mark)

🄔 For show questions, you need to state your answer to a greater number of significant figures than the data given in the question.

(d) Using the value you have determined from (c) and the other data from the figure, calculate the wavelength of the laser light.

(4 marks)

🄔 First you will need to calculate the angle θ using trigonometry.

(e) Suggest one change that you could make to the experiment to reduce the uncertainty in your value for the wavelength.

(2 marks)

🄔 A-grade students will realise that for 2 marks they will need to suggest a change and then describe its effect on the uncertainty.

Student answer

(a) Any three points from:

- The pattern consists of maxima separated by regions of zero intensity ✓.
- The central maximum is the brightest ✓.
- The pattern is symmetrical about central maximum ✓.
- The intensity of the maxima decreases with 'order' number ✓.
- The maxima are equally spaced ✓.
- The maxima are much narrower than spaces between the maxima ✓.

(Maximum 3 marks)

🄔 An A-grade answer will consist of three different points. A C-grade student may well state the same point twice but in different words.

(b) At the maxima, constructive interference *or* the waves add *or* the waves superimpose in-phase *or* the path difference is a whole number of wavelengths (any one of these ✓).

and

Between the maxima, destructive interference *or* the waves cancel *or* the waves in in antiphase *or* the path difference is an odd number of half wavelengths (any one of these ✓).

🄔 You can have any of these answers. However, you will need to explain the maxima and minima in the same way.

(c) $\dfrac{1 \times 10^{-3} \text{ m}}{150}$ ✓ $(= 6.67 \times 10^{-6} \text{ m} \approx 7 \,\mu\text{m})$

(d) $\tan\theta = \dfrac{120 \times 10^{-3} \text{ m}}{1.3 \text{ m}}$ ✓ $\Rightarrow \theta = 5.3°$ ✓

Using the transmission diffraction grating equation:

$\lambda = 6.67 \times 10^{-6} \text{ m} \times \sin 5.3° = 6.2 \times 10^{-7} \text{ m}$ ✓

ⓔ In this case, no unit was given in the question. This means that the correct unit is required as part of the answer. Error carried forward will operate for part (d).

(e) Any one of (1 mark for the suggestion and 1 for the explanation):

- More grating lines per mm ✓. This leads to a larger spacing between maxima to measure ✓.

    or

- Move screen further away from the diffraction grating ✓. This produces a smaller percentage error in the measured distances ✓.

    or

- Measure values using a higher diffraction order ✓. This reduces the percentage error in the measured distances ✓.

Question 15

A university physics department is investigating the behaviour of waves at sea and their effect on the coastline. A large tank is set up with a wave-generating machine at one end. The motion of the machine generates progressive water waves that travel from A to D across water of varying depth (see Figure 5).

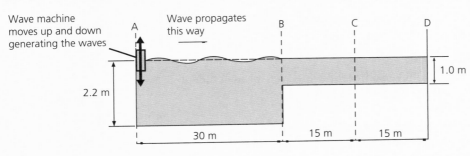

Figure 5

(a) The wave machine oscillates up and down once every 2.5 s. Calculate the frequency of the waves in Hz.

(1 mark)

(b) The waves produced by the wave machine travel from A to B in 6.5 s. Calculate the speed of the waves between A and B.

(1 mark)

(c) Calculate the wavelength of the waves between A and B.

(1 mark)

ⓔ C-grade students are expected to be able to successfully complete questions (a) to (c) relatively easily.

(d) The physicists have found that the speed of the waves at the surface is related to the depth of the water through the equation $c = \sqrt{g \times d}$, where c is the speed of the wave (in m s^{-1}), d is the depth of the water (in m) and g is the acceleration due to gravity, ($9.81\,\text{m s}^{-2}$). Use this equation to determine the speed of the water waves at C.   (1 mark)

ⓔ This question requires you to read the figure correctly and interpret the data successfully. Make sure that you extract the information for the correct point (C).

(e) Explain why refraction occurs at the point B in the tank.   (2 marks)

ⓔ A-grade students will be able to use the correct equation and apply it to this particular boundary related to water waves, as the datasheet equation relates to light.

(f) Calculate the refractive index of the boundary at B.   (2 marks)

ⓔ There are two ways to do this: calculate the speeds in both shallow and deep water or substitute the equation $c = \sqrt{g \times d}$ into the equation for refractive index.

Student answer

(a) $\text{frequency} = \dfrac{1}{\text{time period}}$; time period $= 2.5\,\text{s} \Rightarrow f = \dfrac{1}{2.5\,\text{s}} = 0.40\,\text{Hz}$ ✓

ⓔ The data are given to 2 significant figures, as should the answer.

(b) $\text{speed} = \dfrac{\text{distance travelled}}{\text{time taken}} = \dfrac{30\,\text{m}}{6.5\,\text{s}} = 4.6\,\text{m s}^{-1}$ ✓

(c) wave speed = frequency × wavelength $\Rightarrow \lambda = \dfrac{c}{f} = \dfrac{4.6\,\text{m s}^{-1}}{0.40\,\text{Hz}} = 11.5\,\text{m}$ ✓

ⓔ Error carried forward will apply for part (c).

ⓔ The equations for speed and wave speed (in b and c) must be recalled from GCSE knowledge — they are not on the datasheet.

(d) $c = \sqrt{g \times d} = \sqrt{(9.81\,\text{m s}^{-2} \times 1.0\,\text{m})} = 3.1\,\text{m s}^{-1}$ ✓

(e) Refraction occurs at the boundary between the deep and the shallow water because the waves change speed (high to low) ✓. This causes a reduction in the wavelength of the wave in the shallow water ✓.

ⓔ A common misconception is that refraction always involves a change of direction. When waves hit a boundary in the direction of the normal, there is only a change of speed and therefore wavelength.

(f) $n = \dfrac{c_1}{c_2} = \dfrac{\sqrt{gd_1}}{\sqrt{gd_2}}$ ✓ $= \sqrt{\dfrac{d_1}{d_2}} = \sqrt{\dfrac{2.2\,\text{m}}{1.0\,\text{m}}} = 1.48$ ✓

ⓔ Remember, refractive index is the ratio of the speed of the wave in the incident medium to the speed of the wave in the refracted medium.

Question 16

The de Broglie equation describes the wavelength of particles on the quantum scale:

$$\lambda = \frac{h}{mv}$$

where m is the mass of the particle and v is the speed at which the particle is travelling.

(a) Calculate the wavelength associated with an electron, of mass 9.11×10^{-31} kg travelling at 2.2×10^6 m s^{-1}. (2 marks)

(b) A particle accelerator produces a thin beam of these electrons, which hits a thin foil target with an interatomic spacing of 4.1×10^{-10} m, producing a diffraction pattern on a detector screen situated on the other side of the foil target. Suggest reasons why a diffraction pattern is produced. (2 marks)

ⓔ This question is about how electrons can behave as waves.

> **Student answer**
>
> (a) $\lambda = \dfrac{h}{mv} \Rightarrow \dfrac{6.6 \times 10^{-34} \text{ J s}}{9.11 \times 10^{-31} \text{ kg} \times 2.2 \times 10^6 \text{ m s}^{-1}}$ ✓ $= 3.2 \times 10^{-10}$ m ✓

ⓔ This is a straightforward calculation. The equation is given in the question, so you have to substitute the numbers correctly and calculate the answer.

> (b) The wavelengths of the electrons and spacing are similar, so $\lambda \approx d$ ✓.
>
> This means that the rows of atoms act as a grating *or* the atoms behave like Young's slits *or* constructive and destructive interference occurs *or* the electrons diffract, producing interference ✓.

ⓔ This is an A-grade discriminator question. Most C-grade students will point out that $\lambda \approx d$, but writing a coherent explanation is more challenging.

Question 17

10% of all naturally occurring magnesium consists of atoms of magnesium-25, $^{25}_{12}$Mg. Magnesium-25 forms ions by the loss of two electrons.

mass of proton = 1.673×10^{-27} kg
mass of neutron = 1.675×10^{-27} kg
mass of electron = 9.11×10^{-31} kg

ⓔ This question requires you to interpret an ^A_ZX notation and be able to calculate specific charge.

(a) State the number of protons, neutrons and electrons in an ion of magnesium-25. (3 marks)

ⓔ A straightforward C-grade target question.

(b) Calculate the charge of a magnesium-25 ion. (1 mark)

e This is an easy question to get wrong, primarily due to choosing the wrong number to multiply by e. Common errors involve multiplying e by 25 or 23.

(c) Calculate the specific charge of the magnesium-25 ion. (2 marks)

(d) Magnesium has two other stable, naturally occurring, isotopes: magnesium-24 and magnesium-25. Describe the similarities and the differences in the atomic structure of the three naturally occurring isotopes. (3 marks)

Student answer

(a) $^{25}_{12}Mg$; 12 protons ✓, (25 – 12) = 13 neutrons ✓; 10 electrons ✓ (2 electrons have been taken from the neutral atom).

(b) Atoms are neutral. The ion must have a relative charge of +2e, so $(2 \times (+1.6 \times 10^{-19}\,C)) = +3.2 \times 10^{-19}\,C$ ✓

(c) specific charge of a particle $= \dfrac{\text{charge of the particle}}{\text{mass of the particle}} = \dfrac{Q}{m}$

$$\frac{Q}{m} = \frac{+3.2 \times 10^{-19}\,C}{\left(12 \times 1.673 \times 10^{-27}\,kg\right) + \left(13 \times 1.675 \times 10^{-27}\,kg\right) + \left(10 \times 9.11 \times 10^{-31}\,kg\right)} \checkmark$$

$= +3.2 \times 10^{-20}\,C\,kg^{-1}$ ✓

e Error carried forward will apply for part (c).

e A-grade students will remember to include the mass of the electrons in the calculation of the mass of the ion (even though their mass is not significant). Calculation of specific charge is a common A-grade discriminator.

(d) The three isotopes have the same number of protons (12) ✓ and the same number of electrons (12) ✓ but they have different numbers of neutrons (12, 13 and 14) ✓.

Question 18

A group of students take measurements of the wavelength of a laser using a diffraction grating. A bar chart of their results is shown in Figure 6.

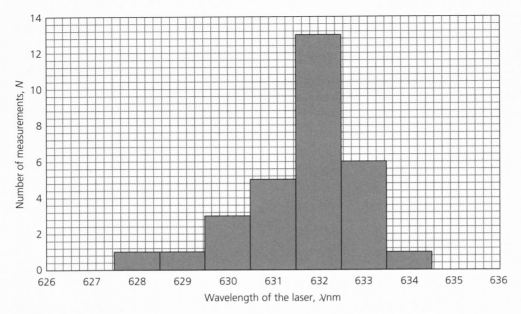

Figure 6

Showing your working clearly, state an estimate of the wavelength of the laser together with an uncertainty of the measurement. Use a suitable number of significant figures.

(3 marks)

ⓔ This question requires you to determine an uncertainty and state an answer to a suitable number of significant figures.

Student answer

The mean value of the wavelength is calculated by:

$$\frac{(628 \times 1)+(629 \times 1)+(630 \times 3)+(631 \times 5)+(632 \times 13)+(633 \times 6)+(634 \times 1)}{30}\text{nm}$$

$$= 632\,\text{nm} \checkmark$$

All the data are three significant figures.

$$\text{uncertainty} = \pm \frac{634-628}{2} \checkmark = \pm 3\,\text{nm}$$

$$\lambda = 632 \pm 3\,\text{nm} \checkmark$$

ⓔ C-grade students generally calculate the mean or state the mode (but this requires an explanation as to why the mode is taken). A-grade students have little difficulty determining the uncertainty using half the spread of the data.

Question 19

White light can reflect off a thin film of soap solution floating on the surface of water. When the reflection is viewed from different angles, different colours can be seen across the soap film. Figure 7 shows an observer looking at two rays of light partially reflected off the upper and lower surfaces of the soap film.

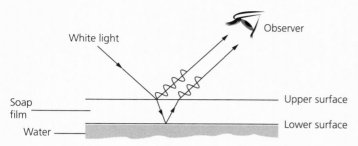

Figure 7

When observing in this direction, the film appears orangey-red in colour due to the absence of blue wavelengths. Explain why this happens. (4 marks)

ⓔ This question requires you to relate wavelength to path difference.

Student answer

At this angle there must be destructive superposition of blue light ✓.

This must happen because the waves arrive at the eye in antiphase ✓. The path difference for the blue light must be an odd number of half wavelengths ✓.

The other colours are still there as they are not in antiphase ✓.

ⓔ An A-grade student would answer this question fully using a well-structured, coherent answer similar to above. A C-grade student would tend to get the first marking point about the destructive interference but would not add the subsequent qualifiers and detail.

Question 20

The incomplete table below gives some features of particles. Identify the values (a), (b), (c) and (d). (4 marks)

Particle	Baryon number, B	Lepton number, L	Relative charge, Q	Quark structure
Proton	+1	0	+1	(a)
Pion	(b)	0	+1	$u\bar{d}$
Positron	0	(c)	+1	None
Electron neutrino	0	+1	(d)	None

ⓔ This is a straightforward question requiring simple recall of values.

Student answer

(a) uud ✓

(b) 0 ✓

(c) –1 ✓

(d) 0 ✓

ⓔ The datasheet contains information regarding the values of Q, B (and strangeness, S) for the u, d and s quarks.

■ Test paper 2

Multiple-choice questions

Question 1

Table 1 below summarises the charge on three quarks.

	Quark		
	u	d	s
Charge, Q	$+\frac{2}{3}$	$-\frac{1}{3}$	$-\frac{1}{3}$
Baryon number, B	$+\frac{1}{3}$	$+\frac{1}{3}$	$+\frac{1}{3}$

Table 1

What is the charge Q and the baryon number B of the Λ baryon (uds)?

	Charge, Q	Baryon number, B
A	+1	+1
B	0	+1
C	+1	−1
D	0	−1

Question 2

What are the fundamental SI units of force?

A $kg\,m\,s^{-1}$

B $kg\,m^{-1}\,s^{-2}$

C $kg\,m\,s^{-2}$

D $kg\,m^{2}\,s^{2}$

Question 3

The Feynman diagram (Figure 1) illustrates the decay of the d quark during beta plus decay.

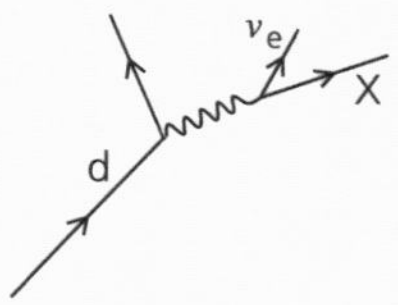

Figure 1

Identify the particle labelled X on the Feynman diagram.

A   u quark

B   W+ exchange particle

C   Electron

D   Positron

Question 4

Here is a list of numbers:

A   0.2

B   2

C   22

D   220

Which number is the best estimate for the work function ϕ of potassium, in eV?

Question 5

Look at the following list of possible graphs that could be drawn using data taken from an experiment to measure the frequency f of the first harmonic standing wave on a string for different tensions T. Which graph would produce a straight line through the origin?

A f against T

B f^2 against T

C f against T^2

D $1/f$ against T^2

Question 6

A ray of light travelling through air is incident on an air–glass boundary. The refractive index of the glass is 1.5 and the angle of incidence to the boundary normal is 36°. The angle of refraction is:

A   23°

B   62°

C   33°

D   36°

Question 7

In an experiment to measure the interference of a coherent monochromatic light source of wavelength 575 nm through a pair of double slits, eight interference maximum fringes occupy a distance of 1.5 cm on a screen 4.2 m away from the slits. The separation of the two slits is:

A   1.3 mm

B   1.6 mm

C   0.16 mm

D   0.13 mm

Question 8

Figure 2 shows the four lowest energy levels of hydrogen. Which arrow represents the energy level change involved with the emission of a photon of energy 1.9 eV?

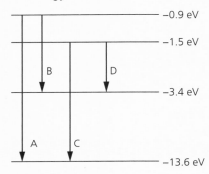

Figure 2

A   B   C   D

Question 9

What is the de Broglie wavelength of positrons of mass 9.11×10^{-31} kg travelling at $0.25\,c$ inside a particle accelerator?

A   2.4 pm

B   9.7 pm

C   2.4 nm

D   9.7 nm

Question 10

Which of the following is not a property of longitudinal waves?

A   Refraction

B   Superposition

C   Polarisation

D   Diffraction

Answers to multiple-choice questions

1 B Addition of the charges gives $Q = 0$; addition of the baryon numbers gives $B = +1$.

2 C $F = ma$, so $kg \times m\,s^{-2} = kg\,m\,s^{-2}$.

3 D During beta plus decay, the d quark decays into a u quark and a W^+ exchange particle, which then decays into an electron neutrino and a positron.

4 B Metals have work functions of a few electronvolts.

5 B $f = \dfrac{1}{2l}\sqrt{\dfrac{T}{\mu}}$ so $f^2 \propto T$.

6 A $n_1 \sin\theta_1 = n_2 \sin\theta_2$; refractive index of air is ≈ 1, so $\theta_2 = 23°$.

7 A $w = \dfrac{\lambda D}{s} \Rightarrow s = \dfrac{\lambda D}{w} = \dfrac{575 \times 10^{-9}\,m \times 4.2\,m}{1.5 \times 10^{-2}\,m/8}$

8 D $hf = E_1 - E_2 = -1.5 - (-3.4)$

9 B $\lambda = \dfrac{h}{mv} = \dfrac{6.6 \times 10^{-34}\,J\,s}{9.11 \times 10^{-31}\,kg \times 0.25 \times 3 \times 10^8\,m\,s^{-1}}$

10 C Polarisation is a property of transverse waves, but not longitudinal waves.

Short and longer answer questions

Question 11

A student was using an old analogue voltmeter to measure the pd across a LED as photons were just being emitted by the LED (called the gating voltage). She repeated the test 25 times, recorded her results and plotted a graph illustrating these data, as shown in Figure 3.

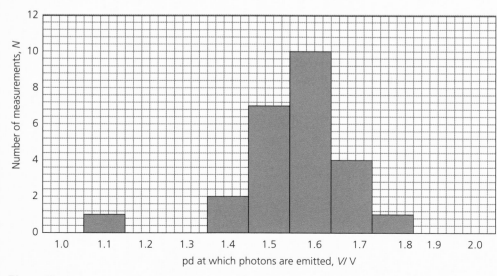

Figure 3

(a) She had a discussion with another student in the class about whether to ignore the result at 1.1 V. Suggest a possible reason for this low result. (1 mark)

ⓔ You may have several possible reasons. Pick the one that gives an answer that is most likely.

(b) She ignored the result at 1.1 V. Estimate the gating voltage of the LED and determine a suitable uncertainty for your estimate. State your answers to a sensible number of significant figures. (3 marks)

ⓔ You will need to remember the method of calculating uncertainty from a spread of data.

Student answer

(a) There are several possible reasons for the low result: a measurement or recording human error, which is the most likely as the student has to judge by eye when the LED is just on; there may be a fault within the voltmeter; background light in the room; poor connections within the circuit. (Any one ✓)

(b) The mean average of the measurements (not including 1.1 V) is:

$$\frac{(1.4\times2)+(1.5\times7)+(1.6\times10)+(1.7\times4)+(1.8\times1)\ V}{24}=1.6\ V$$

The range of the measurements is (1.8 − 1.4) = 0.4 V ✓

Uncertainty = 0.4 V/2 = 0.2 V ✓

Value = 1.6 ± 0.2 V ✓

ⓔ A C-grade student is likely to be able to calculate the mean (although the median and mode would be accepted as answers, provided that there was a suitable explanation). A-grade students would be able to calculate the uncertainty using the spread of the data.

Question 12

Figure 4 shows part of the energy level diagram of mercury. The energy levels are given in electronvolts. One of the photons emitted by a mercury lamp is an ultraviolet photon with a frequency of 1.2×10^{15} Hz.

−3.8 eV ————————————

−5.0 eV ————————————
−5.5 eV ————————————

−10.4 eV ————————————

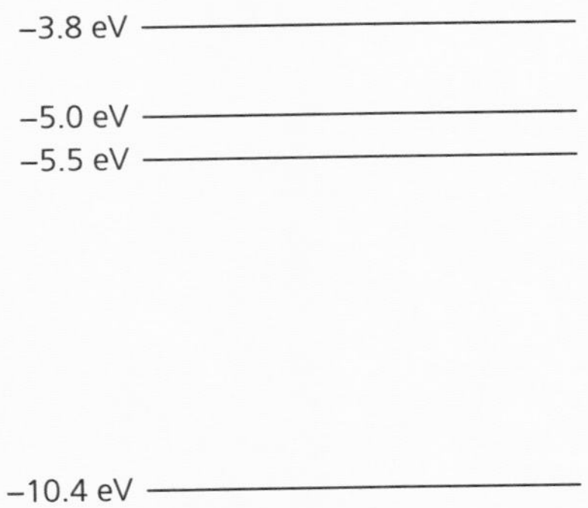

Figure 4

(a) Show that the energy of the ultraviolet photons is about 5 eV.   (2 marks)

(b) On a copy of the diagram, draw an arrow showing the energy level transition that produces the emission of this ultraviolet photon.   (1 mark)

(c) Three possible photons are emitted by energy level transitions from the −3.8 eV energy level. Calculate the longest wavelength of these three photons.   (3 marks)

ⓔ This question is about energy levels. You need to remember that the energy of emitted photons is equal to the difference between two energy levels *and* that energy levels are negative.

Student answer

(a) $E = hf \Rightarrow E = 6.6 \times 10^{-34}\ \text{J s} \times 1.2 \times 10^{15}\ \text{Hz} = 7.92 \times 10^{-19}\ \text{J}$ ✓

$$E\ (\text{eV}) = \frac{E\ (\text{J})}{1.6 \times 10^{-19}\ \text{J eV}^{-1}} = 4.95\ \text{eV} \approx 5\ \text{eV} \text{ ✓}$$

(b)
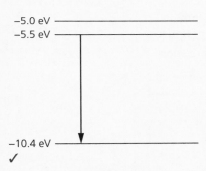
✓

(c) The longest wavelength emission will be from the lowest of these energy transitions so (−3.8 eV to −5.0 V) = 1.2 eV ✓

($= 1.2 \times 1.6 \times 10^{-19}\ \text{J eV}^{-1} = 1.9 \times 10^{-19}\ \text{J}$) ✓

$$E = hf = \frac{hc}{\lambda} \Rightarrow \lambda = \frac{hc}{E} = \frac{6.6 \times 10^{-34}\ \text{J s} \times 3 \times 10^{6}\ \text{m s}^{-1}}{1.9 \times 10^{-19}\ \text{J}} = 1.0 \times 10^{-6}\ \text{m} \text{ ✓}$$

ⓔ Error carried forward will apply for the last part of (c).

ⓔ C-grade students should be able to calculate the wavelength, but they may not convert the energies into joules and they may not identify the longest wavelength correctly.

Question 13

Figure 5 illustrates a photon being converted into an electron and a positron close by to a large nucleus.

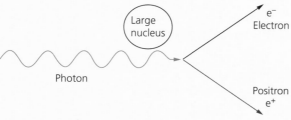

Figure 5

(a) State the name of this process. (1 mark)

(b) The electron has the following properties:

■ rest-mass energy, $m = 0.510999\,\text{MeV}$
■ lepton number, $L = +1$
■ baryon number, $B = 0$
■ (relative) charge, $Q = -1.6 \times 10^{-19}\,\text{C}$

State the corresponding values for the positron. (4 marks)

(c) Calculate the minimum energy of the photon in joules to undergo this process. State your answer to an appropriate number of significant figures. (3 marks)

(d) Describe what is likely to happen to the positron shortly after its creation. (2 marks)

ⓔ You need to know (learn) the mechanisms of both pair production and annihilation.

Student answer

(a) Pair production ✓.

(b) Positron values are:

■ rest-mass energy, $m = 0.510999\,\text{MeV}$ ✓
■ lepton number, $L = -1$ ✓
■ baryon number, $B = 0$ ✓
■ charge, $Q = +1.6 \times 10^{-19}\,\text{C}$ ✓

ⓔ Although this is a straightforward question, the rest-mass energy is stated rather than the more common 'mass'. You must remember that both mass and rest-mass energy are the same for particles and antiparticles.

(c) The energy, E, of the photon must be (rest-mass energy of electron + rest-mass energy of positron):

$E\,(\text{eV}) = 2 \times 0.510999\,\text{MeV} = 1.021998$ ✓ MeV

$E\,(\text{J}) = 1.021998 \times 10^{6}\,\text{eV} \times 1.60 \times 10^{-19}\,\text{J\,eV}^{-1} = 1.64 \times 10^{-13}\,\text{J}$ ✓

(The answer must be to 3 significant figures as e is given to 3 sf ✓)

ⓔ A-grade students avoid the most common error in this question of failing to convert the energy in MeV into joules.

(d) The positron is likely to immediately meet another electron and undergo annihilation ✓, producing a pair of gamma photons ✓.

ⓔ C-grade students will not give the emission of a pair of gamma rays as part of their answer.

Question 14

During electron capture a quark changes its identity.

ⓔ 'Identity' in this case means what type of quark it is.

(a) Explain what is meant by electron capture.   (4 marks)

(b) Draw a Feynman diagram for electron capture.   (3 marks)

Student answer

(a) An (atomic/orbital/shell) electron ✓ interacts with a proton ✓ in the nucleus (via the weak interaction) ✓ forming a neutron and an electron neutrino ✓.

ⓔ You could also describe this decay in terms of a change in quark identity (u to d). C-grade students may just state 'the capture of an electron by a proton' without giving the outcome of the process.

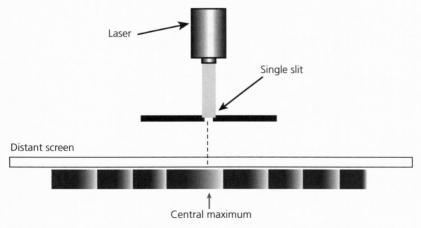

Question 15

A laser emits a ray of monochromatic, coherent blue light that travels through a single slit which is much narrower than the wavelength of the light. A diffraction pattern is produced on a distant screen, as shown in Figure 6.

Laser

Single slit

Distant screen

Central maximum

Figure 6

(a) Describe how the pattern observed on the screen would change if a longer wavelength red light laser were used. (2 marks)

(b) State *two* ways in which the appearance of the diffraction maxima fringes would change if the slit was made *narrower*. (2 marks)

(c) Explain why the following precautions would make this experiment safer:

   (i) Reducing the intensity of the laser beam by putting a neutral density filter in line with the laser. (1 mark)

   (ii) Displaying the laser safety sign, as shown in Figure 7, on the door into the laboratory. (1 mark)

LASER RADIATION

Figure 7

(d) The laser is replaced with a projector bulb producing a beam of white light. Describe how the appearance of the diffraction pattern would change. (3 marks)

ⓔ This is a common question type. Single slit diffraction, Young's double slits and diffraction gratings all involve variables such as wavelength and slit width. When these variables are changed, the effect on the diffraction patterns is the same.

Student answer

(a) The diffraction maxima fringes would be further apart ✓.

The diffraction maxima fringes would be wider ✓.

(b) The diffraction maxima fringes would be wider and have increased separation ✓.

The diffraction maxima fringes would be dimmer (less intensity) ✓.

(c) (i) Reducing the intensity would reduce accidental damage to the retina if the beam were to enter the eye ✓.

   (ii) Displaying the sign would increase the awareness of people entering the laboratory, meaning that they would be less likely to accidentally get their eye into the path of the laser beam ✓.

(d) Any three ✓✓✓ from:
   ■ There would be a central white maxima fringe.
   ■ Each maxima fringe would be composed of a spectrum of colours.
   ■ The violet (or blue) colours would be closest to the central maxima fringe and the red colours would be furthest away.
   ■ The maxima fringes would be wider *or* the dark minimum fringes would be narrower.

ⓔ This type of question is generally done well by A-grade students, who are able to explain their answers more coherently. To help with this, try to write more, shorter sentences, or answer as bullet points.

Question 16

During an experiment to investigate the photoelectric effect, a student shone light of different frequencies onto a sodium photocell. She measured the maximum kinetic energy of the photoelectrons emitted. Her results are shown in Table 2.

Frequency of light photons, f/× 10^{15} Hz	Maximum kinetic energy of emitted photoelectrons, E_{kmax}/× 10^{-19} J
2.0	9.6
1.5	6.4
1.0	3.0
0.5	0.0

Table 2

(a) Using a copy of the graph axes in Figure 8 as a template, plot a graph of maximum kinetic energy (y-axis) against frequency of photons (x-axis).   (2 marks)

ⓔ If you are given a blank graph on your examination paper you need to use as much of the area of the graph paper as possible.

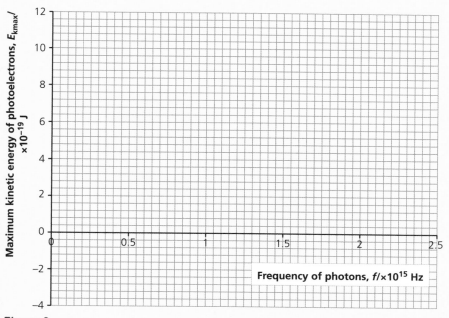

Figure 8

(b) Plot a best-fit line.   (1 mark)

The photoelectric effect equation is:

$hf = E_{kmax} + \phi$

This can be rearranged to give:

$E_{kmax} = hf - \phi$

(c) Use your best-fit line to determine:

    (i) the Planck constant   (2 marks)

    (ii) the work function of sodium   (2 marks)

Student answer

(a) A suitable scale, where the graph fits most of the graph paper ✓.

Correctly plotting the points ✓.

(b)

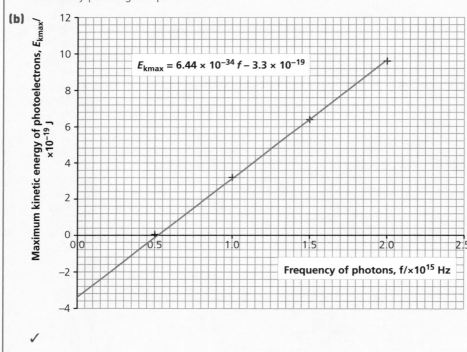

$E_{kmax} = 6.44 \times 10^{-34} f - 3.3 \times 10^{-19}$

✓

(c) (i) The gradient of the line is the Planck constant ✓. Excel gives this to be 6.44×10^{-34} J s ✓.

    (ii) The work function is the y-intercept of the graph ✓. Excel gives this to be 3.3×10^{-19} J ✓.

---

e A common mistake on questions of this type is to take the numerical value of the intercept from the wrong axis. In this case you need the y-axis intercept, which is below the x-axis.

e C-grade students are expected to be able to draw the graph correctly without error. A-grade students are able to calculate the gradients using the standard form values and correctly identify the work function as the y-axis intercept.

Question 17

A student places a microwave transmitter in front of a metal sheet reflector producing a standing wave pattern. She then moves a microwave receiver between the transmitter and the reflector until the receiver detects a maximum *amplitude* signal, as shown in Figure 9.

Figure 9

The reflector is now moved a distance x, slowly to the right. The amplitude of the signal detected is now reduced to a minimum, as shown in Figure 10.

Figure 10

ℯ This question is about standing waves. These involve nodes and antinodes.

(a) The microwaves are emitted at a frequency of 9.5 GHz. Calculate the wavelength of the microwaves. (2 marks)

(b) Use your value from (a) to calculate the distance x. (2 marks)

(c) The student observes that there is an antinode immediately next to the transmitter and another antinode immediately next to the reflector and that there are two nodes between the emitter and the reflector. Draw the standing wave pattern between the emitter and the reflector. (2 marks)

Student answer

(a) $c = f\lambda \Rightarrow \lambda = \dfrac{c}{f} = \dfrac{3 \times 10^8 \, \text{ms}^{-1}}{9.5 \times 10^9 \, \text{Hz}} = 0.032 \, \text{m}$ ✓

(b) Distance x must be $\lambda/4$ as the receiver moves from an antinode to a node. ✓
So $x = 4 \times 0.032 \, \text{m} = 0.128 \, \text{m}$ (0.13 m to 2 sf) ✓

ℯ A-grade students correctly identify x to be $\lambda/4$. Error carried forward will apply in this question, and a correct value of x calculated from the wrong assertion about λ will score 1 mark.

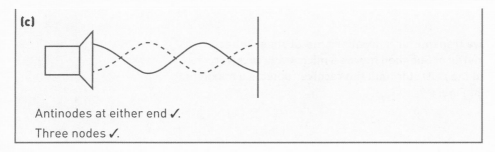

Antinodes at either end ✓.

Three nodes ✓.

Question 18

A beam of red laser light was shone into a pair of two identical triangular glass acrylic (a type of plastic) prisms placed immediately next to each other, as shown in Figure 11.

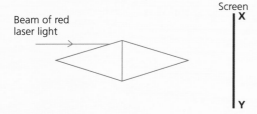

Screen
X

Beam of red
laser light

Y

Figure 11

(a) On a copy of the diagram, complete the path of the beam through the prisms and onto the screen, XY. (2 marks)

(b) The refractive index of acrylic for red light is 1.49. The beam hits the boundary between the air and the acrylic at an angle of 68° to the normal. Calculate the angle of refraction for this boundary. (2 marks)

ⓔ This question requires you to apply Snell's law in a more unusual context. Remember, refractive index varies with wavelength.

(c) The red laser is now replaced with a similar blue laser. The blue beam enters the prisms at the same point and in the same direction as the red laser beam. The refractive index of acrylic is 1.50 for blue light. Explain how the position of the beam on the screen changes. (2 marks)

(d) The light refracts across the boundary because the speed of light changes as the light travels from the air into the acrylic. Use the data to calculate the speed of the blue light in acrylic. (1 mark)

Student answer

(a)

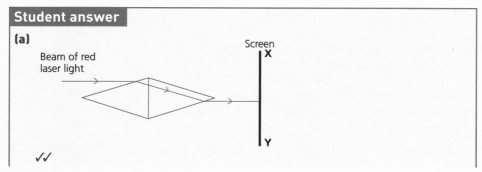

Beam of red
laser light

Screen
X

Y

✓✓

(b)
$$n_1 \sin i_1 = n_2 \sin i_2 \Rightarrow i_2 = \sin^{-1}\left(\frac{n_1}{n_2}\sin i_1\right) \checkmark = \sin^{-1}\left(\frac{1}{1.49}\sin 68°\right) = 38° \checkmark$$

(c) Refraction is dependent on the refractive index — different colours have different refractive indices ✓. The angle of refraction is less in this case so the beam moves towards point Y on the screen ✓.

@ A C-grade student is likely to describe the change in position of the beam, but may well not explain why the beam position moves.

(d) refractive index $= \dfrac{c_1}{c_2} = \dfrac{3 \times 10^8 \, m\,s^{-1}}{c_2} \Rightarrow c_2 = \dfrac{3 \times 10^8 \, m\,s^{-1}}{1.50} = 2 \times 10^8 \, m\,s^{-1}.$ ✓

Question 19

During radioactive emission of α, β^- and β^+ particles, unstable nuclei decay. Complete nuclear equations for the following decays:

(a) Gold-172 $\left(^{172}_{79}Au\right)$ via alpha decay to an isotope of iridium (Ir). (2 marks)

(b) Carbon-14 $\left(^{14}_{6}C\right)$ via beta minus decay to an isotope of nitrogen (N). (3 marks)

(c) Carbon-11 $\left(^{11}_{6}C\right)$ via beta plus decay to an isotope of boron (B). (3 marks)

@ Questions involving decay equations require you to apply the $^A_Z X$-notation and the laws of conservation of baryon and lepton number.

Student answer

(a) $^{172}_{79}Au \rightarrow \,^{172}_{77}Ir \checkmark + \,^4_2He \checkmark$

(b) $^{14}_{6}C \rightarrow \,^{14}_{7}N \checkmark + \,^0_{-1}e \checkmark + \overline{\upsilon}_e \checkmark$

(c) $^{11}_{6}C \rightarrow \,^{11}_{5}B \checkmark + \,^0_{+1}e \checkmark + \upsilon_e \checkmark$

@ You need to learn the changes to the values of A and Z during radioactive decay. A-grade students will remember to conserve lepton number as well as baryon number, particularly during the weak interactions.

Question 20

During neutron crystallography, a beam of neutrons from a nuclear reactor hits a powdered sample of crystals and produces a diffraction pattern on a detector screen.

Questions & Answers

(a) Explain why the de Broglie wavelength of the neutrons must be about 1 nm for this application. (2 marks)

(b) The structure of crystals can also be probed using electron diffraction. Give one disadvantage of using neutrons rather than electrons for this purpose. (1 mark)

(c) Calculate the de Broglie wavelength of neutrons with a kinetic energy of 1.1×10^{-22} J emitted by a nuclear reactor. (3 marks)

ⓔ In this question you will first have to calculate the momentum of the neutrons. Remember, the mass of a proton, neutron and electron is given in the datasheet.

(d) In one such experiment, the neutrons in (c) produced an $n = 1$ diffraction maximum at an angle of 53° to the normal of a crystal sample. Use the diffraction grating equation, $n\lambda = d \sin\theta$, to calculate the spacing of the atoms in the crystal. (2 marks)

Student answer

(a) The spacing of atoms in a crystal is the same order of magnitude as the wavelength of the neutrons ✓.Maximum diffraction occurs when this condition is met ✓.

(b) As neutrons are uncharged, their speed, momentum and therefore wavelength are difficult to control using electric and magnetic fields ✓.

(c) $p = mv$ so $p^2 = m^2 v^2$ and

$$E_k = \frac{1}{2}mv^2 \Rightarrow E_k = \frac{p^2}{2m} \Rightarrow p = \sqrt{2mE_k} \; \checkmark \; =$$

$$\sqrt{2 \times 1.675 \times 10^{-27} \, \text{kg} \times 1.1 \times 10^{-22} \, \text{J}} = 6.1 \times 10^{-25} \, \text{kg m s}^{-1} \; \checkmark$$

$$\lambda = \frac{h}{p} = \frac{6.6 \times 10^{-34} \, \text{J s}}{6.1 \times 10^{-25} \, \text{kg m s}^{-1}} = 1.1 \times 10^{-9} \, \text{m} \; \checkmark$$

ⓔ This is an A-grade discriminator question. Error carried forward will also operate for calculation of the wavelength from an incorrect momentum.

(d) $n\lambda = d\sin\theta \Rightarrow d = \dfrac{1.1 \times 10^{-9} \, \text{m}}{\sin 53°} \; \checkmark = 1.4 \times 10^{-9} \, \text{m} \; \checkmark$

Knowledge check answers

1 Metre; kilogram; second; ampere; kelvin; mole; candela

2 i) ms^{-2}, ii) $kg\,m^2\,s^{-3}$, iii) $kg\,m^2\,A^{-1}\,s^{-3}$

3 5×10^{-6} V; 5 μV

4 9×10^{7} J or 90 MJ

5 Systematic errors affect a set of measurements the same way each time, whereas a random error affects measurements in an unpredictable fashion.

6 The measurements have been carried out by many different physicists across the world over many years and the values are always the same (within experimental uncertainty).

7 $11\,ms^{-1}$ — the least precise measurement has 2 sf.

8 $5.7 \pm 0.8\,ms^{-1}$

9 Four orders of magnitude.

10 Protons and neutrons.

11

$$\text{specific charge of a particle} = \frac{\text{charge of the particle}}{\text{mass of the particle}}$$

$$= \frac{Q}{m}$$

For a deuteron, specific charge = $+5.71 \times 10^{34}\,C\,kg^{-1}$

12 Atoms of the same element with different numbers of neutrons.

13 $^{1}_{1}H$; $^{2}_{1}H$; $^{3}_{1}H$

14 Force that holds nucleons together.

15 $^{226}_{88}Ra \rightarrow {}^{222}_{84}Rn + {}^{4}_{2}He$

16 To account for the apparent violation of the law of conservation of energy.

17 When a particle meets its antiparticle and converts into two gamma ray photons.

18 Two gamma rays are emitted at the same time and travel in opposite directions and are detected by the scanner, which rotates around the patient detecting similar events in different directions.

19 Beta minus emission involves the decay of neutrons and the emission of electrons; beta plus emission involves the decay of protons and emission of positrons.

20 A baryon consists of three quarks whereas a meson consists of a quark–antiquark pair.

21 Zero.

22 Hadrons are made up of quarks, leptons are fundamental particles.

23 Proton = uud; $Q = +\frac{2}{3} + \frac{2}{3} - \frac{1}{3} = +1$

24 $u \Rightarrow d$

25 3.5 eV

26 660 nm

27 2.18×10^{-18} J

28 Excitation — electrons move up to higher energy levels; ionisation — electrons removed from the atom.

29 Particles (electrons) behaving as waves.

30 1.65×10^{-11} m

31 25 kHz

32 4.74×10^{14} Hz

33 180°; π^c

34 $330\,ms^{-1}$

35 Transverse waves can be polarised, longitudinal waves cannot.

36 1100 Hz

37 16 Hz

38 7.5 cm; path difference = 5 × wavelength/2, therefore destructive superposition occurs.

39 530 nm

40 w is proportional to λ, so the different colours have different fringe spacings producing spectral fringes.

41 306 lines per mm

42 1.5

43 30°

44 Yes, the critical angle for this boundary is 60°, so the angle is lower than the critical angle.

45 The cladding has a lower refractive index than the material that the fibre is made of ensuring that the signal inside the fibre undergoes total internal reflection.

Answers to required practicals

Required practical 1

The error bars ($\pm 10\,\text{Hz}^2$) are too small to be seen on the graph.

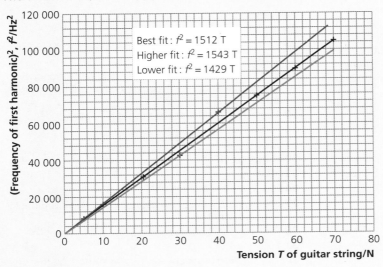

The gradient of the line is equal to $1/4l^2\mu$ so:

$$\mu = \frac{1}{4l^2 \times \text{gradient}}$$

The gradients shown on the graph have been determined using Excel. Inserting these values into the equation:

$\mu_{\text{bestfit}} = 4.0 \times 10^{-4}\,\text{kg m}^{-1}$

$\mu_{\text{higher gradient}} = 3.9 \times 10^{-4}\,\text{kg m}^{-1}$

$\mu_{\text{lower gradient}} = 4.3 \times 10^{-4}\,\text{kg m}^{-1}$

$\mu = (4.0 \pm 0.2) \times 10^{-4}\,\text{kg m}^{-1}$

Required practical 2

For the double slit:

$$w = \frac{\lambda D}{s}$$

and from the diffraction pattern 11 bright fringes occur between 2.5 cm and 5.0 cm, so the fringe spacing w is given by:

$$w = \frac{2.5 \times 10^{-2}\,\text{m}}{11} = 2.27 \times 10^{-3}\,\text{m}$$

Rearranging the formula gives:

$$s = \frac{\lambda D}{w} = \frac{632.8 \times 10^{-9}\,\text{m} \times 2.3\,\text{m}}{2.27 \times 10^{-3}\,\text{m}} = 6.4 \times 10^{-4}\,\text{m}$$

For the diffraction grating with first order diffraction maximum, $n = 1$, and using the small angle approximation:

$$\theta \approx \sin\theta \approx \tan\theta \approx \frac{1.15\,\text{cm}}{230\,\text{cm}} = 5.0 \times 10^{-3}$$

Rearranging the formula gives:

$$d = \frac{n\lambda}{\sin\theta} = \frac{1 \times 632.8 \times 10^{-9}\,\text{m}}{5.0 \times 10^{-3}} = 1.26 \times 10^{-4}\,\text{m} \left(1.3 \times 10^{-4}\,\text{m to 2 sf}\right)$$

Index

Note: **bold** page numbers indicate defined terms.

Index